SpringerBriefs in Electrical and Computer Engineering

More information about this series at http://www.springer.com/series/10059

Linjiun Tsai · Wanjiun Liao

Virtualized Cloud Data Center Networks: Issues in Resource Management

 Springer

Linjiun Tsai
National Taiwan University
Taipei
Taiwan

Wanjiun Liao
National Taiwan University
Taipei
Taiwan

ISSN 2191-8112   ISSN 2191-8120   (electronic)
SpringerBriefs in Electrical and Computer Engineering
ISBN 978-3-319-32630-6   ISBN 978-3-319-32632-0   (eBook)
DOI 10.1007/978-3-319-32632-0

Library of Congress Control Number: 2016936418

Printed on acid-free paper

This Springer imprint is published by Springer Nature
The registered company is Springer International Publishing AG Switzerland

Preface

This book introduces several important topics in the management of resources in virtualized cloud data centers. They include consistently provisioning predictable network quality for large-scale cloud services, optimizing resource efficiency while reallocating highly dynamic service demands to VMs, and partitioning hierarchical data center networks into mutually exclusive and collectively exhaustive subnetworks.

To explore these topics, this book further discusses important issues, including (1) reducing hosting cost and reallocation overheads for cloud services, (2) provisioning each service with a network topology that is non-blocking for accommodating arbitrary traffic patterns and isolating each service from other ones while maximizing resource utilization, and (3) finding paths that are link-disjoint and fully available for migrating multiple VMs simultaneously and rapidly.

Solutions which efficiently and effectively allocate VMs to physical servers in data center networks are proposed. Extensive experiment results are included to show that the performance of these solutions is impressive and consistent for cloud data centers of various scales and with various demands.

Contents

Chapter 1
Introduction

1.1 Cloud Computing

Cloud computing lends itself to the processing of large data volumes and time-varying computational demands. Cloud data centers involve substantial computational resources, feature inherently flexible deployment, and deliver significant economic benefit—provided the resources are well utilized while the quality of service is sufficient to attract as many tenants as possible.

Given that they naturally bring economies of scale, research in cloud data centers has received extensive attention in both academia and industry. In large-scale public data centers, there may exist hundreds of thousands of servers, stacked in racks and connected by high-bandwidth hierarchical networks to jointly form a shared resource pool for accommodating multiple cloud tenants from all around the world. The servers are provisioned and released on-demand via a self-service interface at any time, and tenants are normally given the ability to specify the amount of CPU, memory, and storage they require. Commercial data centers usually also offer service-level agreements (SLAs) as a formal contract between a tenant and the operator. The typical SLA includes penalty clauses that spell out monetary compensations for failure to meet agreed critical performance objectives such as downtime and network connectivity.

1.2 Server Virtualization

Virtualization [1] is widely adopted in modern cloud data centers for its agile dynamic server provisioning, application isolation, and efficient and flexible resource management. Through virtualization, multiple instances of applications can be hosted by virtual machines (VMs) and thus separated from the underlying hardware resources. Multiple VMs can be hosted on a single physical server at one

© The Author(s) 2016
L. Tsai and W. Liao, *Virtualized Cloud Data Center Networks:
Issues in Resource Management*, SpringerBriefs in Electrical
and Computer Engineering, DOI 10.1007/978-3-319-32632-0_1

time, as long as their aggregate resource demand does not exceed the server capacity. VMs can be easily migrated [2] from one server to another via network connections. However, without proper scheduling and routing, the migration traffic and workload traffic generated by other services would compete for network bandwidth. The resultant lower transfer rate invariably prolongs the total migration time. Migration may also cause a period of downtime to the migrating VMs, thereby disrupting a number of associated applications or services that need continuous operation or response to requests. Depending on the type of applications and services, unexpected downtime may lead to severe errors or huge revenue losses. For data centers claiming high availability, how to effectively reduce migration overhead when reallocating resources is therefore one key concern, in addition to pursuing high resource utilization.

1.3 Server Consolidation

The resource demands of cloud services are highly dynamic and change over time. Hosting such fluctuating demands, the servers are very likely to be underutilized, but still incur significant operational cost unless the hardware is perfectly energy proportional. To reduce costs from inefficient data center operations and the cost of hosting VMs for tenants, server consolidation techniques have been developed to pack VMs into as few physical servers as possible, as shown in Fig. 1.1. The techniques usually also generate the reallocation schedules for the VMs in response to the changes in their resource demands. Such techniques can be used to consolidate all the servers in a data center or just the servers allocated to a single service.

Fig. 1.1 An example of server consolidation

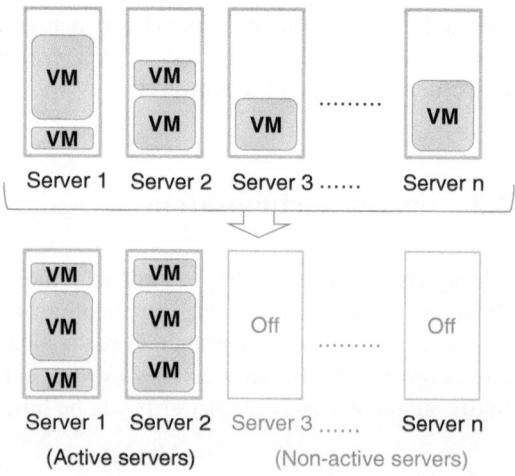

Server consolidation is traditionally modeled as a bin-packing problem (BPP) [3], which aims to minimize the total number of bins to be used. Here, servers (with limited capacity) are modeled as bins and VMs (with resource demand) as items.

Previous studies show that BPP is NP-complete [4] and many good heuristics have been proposed in the literature, such as First-Fit Decreasing (FFD) [5] and First Fit (FF) [6], which guarantee that the number of bins used, respectively, is no more than $1.22\,N + 0.67$ and $1.7\,N + 0.7$, where N is the optimal solution to this problem. However, these existing solutions to BPP may not be directly applicable to server consolidation in cloud data centers. To develop solutions feasible for clouds, it is required to take into account the following factors: (1) the resource demand of VMs is dynamic over time, (2) migrating VMs among physical servers will incur considerable overhead, and (3) the network topology connecting the VMs must meet certain requirements.

1.4 Scheduling of Virtual Machine Reallocation

When making resource reallocation plans that may trigger VM migration, it is necessary to take into account network bandwidth sufficiency between the migration source and destination to ensure the migration can be completed in time. Care must also be taken in selecting migration paths so as to avoid potential mutual interference among multiple migrating VMs. Trade-off is inevitable and how well scheduling mechanisms strike balances between the migration overhead and quality of resource reallocation will significantly impact the predictability of migration time, the performance of data center networks, the quality of cloud services and therefore the revenue of cloud data centers.

The problem may be further exacerbated in cloud data centers that host numerous services with highly dynamic demands, where reallocation may be triggered more frequently because the fragmentation of the resource pool becomes more severe. It is also more difficult to find feasible migration paths on saturated cloud data center networks. To reallocate VMs that continuously communicate with other VMs, it is also necessary to keep the same perceived network quality after communication routing paths are changed. This is especially challenging in cloud data centers with communication-intensive applications.

1.5 Intra-Service Communications

Because of the nature of cloud computing, multiple uncooperative cloud services may coexist in a multi-tenant cloud data center and share the underlying network. The aggregate traffic from the services is highly likely to be dynamic [7], congested [8] or even malicious [9] in an unpredictable manner. Without effective mechanisms for service allocation and traffic routing, the intra-service communication of every service sharing the network may suffer serious delay and even be disrupted. Deploying all the VMs for a service into one single rack to reduce the impact on the shared network is not always a practical or economical solution. This is because such a solution may cause the resources of data centers to be underutilized and fragmented, particularly when the demand of services is highly dynamic and does not fit the capacity of the rack.

For delay-sensitive and communication-intensive applications, such as mobile cloud streaming [10, 11], cloud gaming [12, 13], MapReduce applications [14], scientific computing [15] and Spark applications [16], the problem may become more acute due to their much greater impact on the shared network and much stricter requirements in the quality of intra-service transmissions. Such types of applications usually require all-to-all communications to intensively exchange or shuffle data among distributed nodes. Therefore, network quality becomes the primary bottleneck of their performance. In some cases, the problem remains quite challenging even if the substrate network structure provides high capacity and rich connectivity, or the switches are not oversubscribed. First, all-to-all traffic patterns impose strict topology requirements on allocation. Complete graphs, star graphs or some graphs of high connectivity are required for serving such traffic, which may be between any two servers. In a data center network where the network resource is highly fragmented or partially saturated, such topologies are obviously extremely difficult to allocate, even with significant reallocation cost and time. Second, dynamically reallocating such services without affecting their performance is also extremely challenging. It is required to find reallocation schedules that not only satisfy general migration requirements, such as sufficient residual network bandwidth, but also keep their network topologies logically unchanged.

To host delay-sensitive and communication-intensive applications with network performance guarantees, the network topology and quality (e.g., bandwidth, latency and connectivity) should be consistently guaranteed, thus continuously supporting arbitrary intra-service communication patterns among the distributed compute nodes and providing good predictability of service performance. One of the best approaches is to allocate every service a non-blocking network. Such a network must be isolated from any other service, be available during the entire service lifetime even when some of the compute nodes are reallocated, and support all-to-all communications. This way, it can give each service the illusion of being operated on the data center exclusively.

1.6 Topology-Aware Allocation

For profit-seeking cloud data centers, the question of how to efficiently provision non-blocking topologies for services is a crucial one. It also principally affects the resource utilization of data centers. Different services may request various virtual topologies to connect their VMs, but it is not necessary for data centers to allocate

Fig. 1.2 Different resource efficiencies of non-blocking networks

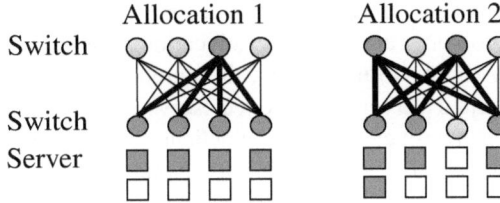

the physical topologies for them in exactly the same form. In fact, keeping such consistency could lead to certain difficulties in optimizing the resources of entire data center networks, especially when such services request physical topologies of high connectivity degrees or even cliques.

For example, consider the deployment of a service which requests a four-vertex clique to serve arbitrary traffic patterns among four VMs on a network with eight switches and eight servers. Suppose that the link capacity is identical to the bandwidth requirement of the VM, so there are at least two feasible methods of allocation, as shown in Fig. 1.2. Allocation 1 uses a star topology, which is clearly non-blocking for any possible intra-service communication patterns, and occupies the minimum number of physical links. Allocation 2, however, shows an inefficient allocation as two more physical links are used to satisfy the same intra-service communication requirements.

Apart from allocating more resources, the star network in Allocation 1 provides better flexibility in reallocation than other complex structures. This is because Allocation 1 involves only one link when reallocating any VM while ensuring topology consistency. Such a property makes it easier for resources to be reallocated in a saturated or fragmented data center network, and thus further affects how well the resource utilization of data center networks could be optimized, particularly when the demands dynamically change over time. However, the question then arises as to how to efficiently allocate every service as a star network. In other words, how to efficiently divide the hierarchical data center networks into a large number of star networks for services and dynamically reallocate those star networks while maintaining high resource utilization? To answer this question, the topology of underlying networks needs to be considered. In this book, we will introduce a solution to tackling this problem.

1.7   Summary

So far, the major issues, challenges and requirements for managing the resources of virtualized cloud data centers have been addressed. The solutions to these problems will be explored in the following chapters. The approach is to divide the problems into two parts. The first one is to allocate VMs for every service into one or multiple virtual servers, and the second one is to allocate virtual servers for all services to physical servers and to determine network links to connect them. Both sub-problems are dynamic allocation problems. This is because the mappings from VMs to virtual servers, the number of required virtual servers, the mapping from virtual servers to physical servers, and the allocation of network links may all change over time. For practical considerations, these mechanisms are designed to be scalable and feasible for cloud data centers of various scales so as to accommodate services of different sizes and dynamic characteristics.

The mechanism for allocating and reallocating VMs on servers is called *Adaptive Fit* [17], which is designed to pack VMs into as few servers as possible. The challenge is not just to simply minimize the number of servers. As the demand of every VM may change over time, it is best to minimize the reallocation overhead by selecting and keeping some VMs on their last hosting server according to an estimated saturation degree.

The mechanism for allocating and reallocating physical servers is based on a framework called *StarCube* [18], which ensures every service is allocated with an isolated non-blocking star network and provides some fundamental properties that allow topology-preserving reallocation. Then, a polynomial-time algorithm will be introduced which performs on-line, on-demand and cost-bounded server allocation and reallocation based on those promising properties of *StarCube*.

References

1. P. Barham et al., Xen and the art of virtualization. ACM SIGOPS Operating Syst. Rev. **37**(5), 164–177 (2003)
2. C. Clark et al., in *Proceedings of the 2nd Conference on Symposium on Networked Systems Design & Implementation*, Live migration of virtual machines, vol. 2 (2005)
3. V.V. Vazirani, *Approximation Algorithms*, Springer Science & Business Media (2002)
4. M.R. Garey, D.S. Johnson, *Computers and intractability: a guide to the theory of NP-completeness* (WH Freeman & Co., San Francisco, 1979)
5. G. Dósa, The tight bound of first fit decreasing bin-packing algorithm is FFD(I) = (11/9)OPT (I) + 6/9, *Combinatorics, Algorithms, Probabilistic and Experimental Methodologies*, Springer Berlin Heidelberg (2007)
6. B. Xia, Z. Tan, Tighter bounds of the first fit algorithm for the bin-packing problem. Discrete Appl. Math. **158**(15), 1668–1675 (2010)
7. Q. He et al., in *Proceedings of the 19th ACM International Symposium on High Performance Distributed Computing*, Case study for running HPC applications in public clouds, (2010)
8. S. Kandula et al., in *Proceedings of the 9th ACM SIGCOMM Conference on Internet Measurement Conference*, The nature of data center traffic: measurements & analysis (2009)
9. T. Ristenpart et al., in *Proceedings of the 16th ACM Conference on Computer and Communications Security*, Hey, you, get off of my cloud: exploring information leakage in third-party compute clouds (2009)
10. C.F. Lai et al., A network and device aware QoS approach for cloud-based mobile streaming. IEEE Trans. on Multimedia **15**(4), 747–757 (2013)
11. X. Wang et al., Cloud-assisted adaptive video streaming and social-aware video prefetching for mobile users. IEEE Wirel. Commun. **20**(3), 72–79 (2013)
12. R. Shea et al., Cloud gaming: architecture and performance. IEEE Network Mag. **27**(4), 16–21 (2013)

13. S.K. Barker, P. Shenoy, in *Proceedings of the first annual ACM Multimedia Systems*, Empirical evaluation of latency-sensitive application performance in the cloud (2010)
14. J. Ekanayake et al., in *IEEE Fourth International Conference on eScience*, MapReduce for data intensive scientific analyses (2008)
15. A. Iosup et al., Performance analysis of cloud computing services for many-tasks scientific computing, *IEEE Trans. on Parallel and Distrib. Syst.* **22**(6), 931–945 (2011)
16. M. Zaharia et al., in *Proceedings of the 2nd USENIX conference on Hot topics in cloud computing*, Spark: cluster computing with working sets (2010)
17. L. Tsai, W. Liao, in *IEEE 1st International Conference on Cloud Networking*, Cost-aware workload consolidation in green cloud datacenter (2012)
18. L. Tsai, W. Liao, StarCube: an on-demand and cost-effective framework for cloud data center networks with performance guarantee, *IEEE Trans. on Cloud Comput.* doi:10.1109/TCC. 2015.2464818

Chapter 2
Allocation of Virtual Machines

In this chapter, we introduce a solution to the problem of cost-effective VM allocation and reallocation. Unlike traditional solutions, which typically reallocate VMs based on a greedy algorithm such as First Fit (each VM is allocated to the first server in which it will fit), Best Fit (each VM is allocated to the active server with the least residual capacity), or Worse Fit (each VM is allocated to the active server with the most residual capacity), the proposed solution strikes a balance between the effectiveness of packing VMs into few servers and the overhead incurred by VM reallocation.

2.1 Problem Formulation

We consider the case where a system (e.g., a cloud service or a cloud data center) is allocated with a number of servers denoted by H and a number of VMs denoted by V. We assume the number of servers is always sufficient to host the total resource requirement of all VMs in the system. Thus, we focus on the consolidation effectiveness and the migration cost incurred by the server consolidation problem.

Further, we assume that VM migration is performed at discrete times. We define the period of time to perform server consolidation as an epoch. Let $T = \{t_1, t_2, \ldots, t_k\}$ denote the set of epochs to perform server consolidation. The placement sequence for VMs in V in each epoch t is then denoted by $F = \{f_t \mid \forall\, t \in T\}$, where f_t is the VM placement at epoch t and defined as a mapping $f_t : V \to H$, which specifies that each VM i, $i \in V$, is allocated to server $f_t(i)$. Note that "$f_t(i) = 0$" denotes that VM i is not allocated. To model the dynamic nature of the resource requirement and the migration cost for each VM over time, we let $R_t = \{r_t(i) \mid \forall\, i \in V\}$ and $C_t = \{c_t(i) \mid \forall\, i \in V\}$ denote the sets of the resource requirement and migration cost, respectively, for all VMs in epoch t.

© The Author(s) 2016
L. Tsai and W. Liao, *Virtualized Cloud Data Center Networks:*
Issues in Resource Management, SpringerBriefs in Electrical
and Computer Engineering, DOI 10.1007/978-3-319-32632-0_2

The capacity of a server is normalized (and simplified) to one, which may correspond to the total resource in terms of CPU, memory, etc. in the server. The resource requirement of each VM varies from 0 to 1. When a VM demands zero resource, it indicates that the VM is temporarily out of the system. Since each server has limited resources, the aggregate resource requirement of VMs on a server must be less than or equal to one. Each server may host multiple VMs with different resource requirements, and each application or service may be distributed to multiple VMs hosted by different servers. A server with zero resource requirements from VMs will not be used to save the hosting cost. We refer to a server that has been allocated VMs as an *active* server.

To jointly consider the consolidation effectiveness and the migration overhead for server consolidation, we define the total cost for VM placement F as the total hosting cost of all active servers plus the total migration cost incurred by VMs. The hosting cost of an active server is simply denoted by a constant E and the total hosting cost for VM placement sequence F is linearly proportional to the number of active servers. To account for revenue loss, we model the downtime caused by migrating a VM as the migration cost for the VM. The downtime could be estimated as in [1] and the revenue loss depends on the contracted service level. Since the downtime is mainly affected by the memory dirty rate (i.e., the rate at which memory pages in the VM are modified) of VM and the network bandwidth [1], the migration cost is considered independent of the resource requirement for each VM.

Let H'_t be a subset of H which is active in epoch t and $|H'_t|$ be the number of servers in H'_t. Let C'_t be the migration cost to consolidate H'_t from epoch t to $t + 1$. H'_t and C'_t are defined as follows, respectively:

$$H'_t = \{f_t(i)|f_t(i) \neq 0, \forall i \in V\}, \forall t \in T$$

$$C'_t = \sum_{i \in V, f_t(i) \neq f_{t+1}(i)} c_t(i), \forall t \in T \backslash \{t_k\}$$

The total cost of F can be expressed as follows:

$$Cost(F) = E \times \sum_{t \in T} |H'_t| + \sum_{t \in T \backslash \{t_k\}} C'_t$$

We study the Total-Cost-Aware Consolidation (TCC) problem. Given $\{H, V, R, C, T, E\}$, a VM placement sequence F is feasible only if the resource constraints for all epochs in T are satisfied. The TCC problem is stated as follows: among all possible feasible VM placements, to find a feasible placement sequence F whose total cost is minimized.

2.2 *Adaptive Fit* **Algorithm**

The TCC problem is NP-Hard, because it is at least as hard as the server consolidation problem. In this section, we present a polynomial-time solution to the problem. The design objective is to generate VM placement sequences F in polynomial time and minimize $Cost(F)$.

Recall that the migration cost results from changing the hosting servers of VMs during the VM migration process. To reduce the total migration cost for all VMs, we attempt to minimize the number of migrations without degrading the effectiveness of consolidation. To achieve this, we try to allocate each VM i in epoch t to the same server hosting the VM in epoch $t-1$, i.e., $f_t(i) = f_{t-1}(i)$. If $f_{t-1}(i)$ does not have enough capacity in epoch t to satisfy the resource requirement for VM i or is currently not active, we then start the remaining procedure based on "saturation degree" estimation. The rationale behind this is described as follows.

Instead of using a greedy method as in existing works, which typically allocate each migrating VM to an active server with available capacity either based on First Fit, Best Fit, or Worse Fit, we define a total cost metric called saturation degree to strike a balance between the two conflicting factors: consolidation effectiveness and migration overhead. For each iteration of allocation process in epoch t, the saturation degree X_t is defined as follows:

$$X_t = \frac{\sum_{i \in V} r_t(i)}{(|H'_t| + 1) \times 1}$$

Since the server capacity is normalized to one in this book, the denominator indicates the total capacity summed over all active servers plus an idle server in epoch t.

During the allocation process, X_t decreases as $|H'_t|$ increases by definition. We define the saturation threshold $u \in [0, 1]$ and say that X_t is low when $X_t \le u$. If X_t is low, the migrating VMs should be allocated to the set of active servers unless there are no active servers that have sufficient capacity to host them. On the other hand, if X_t is large (i.e., $X_t > u$), the mechanism tends to "lower" the total migration cost as follows. One of the idle servers will be turned on to host a VM which cannot be allocated on its "last hosting server" (i.e., $f_{t-1}(i)$ for VM i), even though some of the active servers still have sufficient residual resource to host the VM. It is expected that the active servers with residual resource in epoch t are likely to be used for hosting other VMs which were hosted by them in epoch $t-1$. As such, the total migration cost is minimized.

The process of allocating all VMs in epoch t is then described as follows. In addition, the details of the mechanism are shown in the Appendix.

1. Sort all VMs in V by decreasing order based on their $r_t(i)$.
2. Select VM i with the highest resource requirement among all VMs not yet allocated, i.e.,

$$i \leftarrow \arg \max_j \{r_t(j)|f_t(j) = 0, \forall j \in V\}$$

3. Allocate VM i:

   (i)   If VM i's hosting server at epoch $t-1$, i.e., $f_{t-1}(i)$, is currently active and has sufficient capacity to host VM i with the requirement $r_t(i)$, VM i is allocated to it, i.e., $f_t(i) \leftarrow f_{t-1}(i)$;

   (ii) If VM i's last hosting server $f_{t-1}(i)$ is idle, and there are no active servers which have sufficient residual resource for allocating VM i or the X_t exceeds the saturation threshold u, then VM i is allocated to its last hosting server, namely, $f_t(i) \leftarrow f_{t-1}(i)$;

   (iii) If Cases i and ii do not hold, and X_t exceeds the saturation threshold u or there are no active servers which have sufficient residual resource to host VM i, VM i will be allocated to an idle server;

   (iv) If Cases i, ii and iii do not hold, VM i is allocated to an active server based on Worse-Fit policy.

4. Update the residual capacity of $f_t(i)$ and repeat the procedure for allocating the next VM until all VMs are allocated.

We now illustrate the operation of *Adaptive Fit* with an example where the system is allocating ten VMs, of which the resource requirements are shown in Table 2.1.

The saturation threshold u is set to one. The step-by-step allocation for epoch t is shown in Table 2.2. The row of epoch $t-1$ indicates the last hosting servers (i.e., $f_{t-1}(i)$) of VMs. The rows for epoch t depict the allocation iterations, with allocation sequence from top to bottom. For each VM, the items underlined denote the actual allocated server while the other items denote the candidate servers with sufficient capacity. The indicators L, X, N, A denote that the allocation decision is based on the policies: (1) Use the last hosting server first; (2) Create new server at high saturation; (3) Create new server because that there is no sufficient capacity in active serves; (4) Use active server by Worse Fit allocation, respectively. Note that the total resource requirement of all VMs is 3.06 and the optimal number of servers to use is 4 in this example. In this example, *Adaptive Fit* can achieve a performance quite close to the optimal.

Table 2.1 Resource requirements for VMs

v_i	$r_t(i)$	v_i	$r_t(i)$
4	0.49	10	0.34
8	0.48	1	0.15
3	0.47	7	0.15
2	0.43	6	0.13
5	0.35	9	0.07

Table 2.2 An example of allocating VMs by *Adaptive Fit*

Epoch	Server 1	Server 2	Server 3	Server 4
$t-1$	v_1, v_9	v_6, v_8	v_5, v_{10}	v_2, v_3
				v_4, v_7
t				$v_4\ (L)$
		$v_8\ (L)$		v_8
		v_3		$v_3\ (L)$
	$v_2\ (X)$	v_2		
	$v_5\ (A)$	v_5		
		$v_{10}\ (A)$		
	$v_1\ (L)$	v_1		
		$v_7\ (A)$		
			$v_6\ (N)$	
	$v_9\ (L)$		v_9	

2.3 Time Complexity of *Adaptive Fit*

Theorem 2.1 (Time complexity of Adaptive Fit) *Adaptive Fit is a polynomial-time algorithm with average-case complexity $O(n \log n)$, where n is the number of VMs in the system.*

Proof We examine the time complexity of each part in *Adaptive Fit*. Let m denote the number of active servers in the system. The initial phase requires $O(m \log m)$ and $O(n \log n)$ to initialize A and A' and V', which are implemented as binary search trees. The operations on A and A' can be done in $O(\log m)$. The saturation degree estimation takes $O(1)$ for calculating the denominator based on the counter for the number of servers used while the numerator is static and calculated once per epoch. The rest of the lines in the "for" loop are $O(1)$. Therefore, the main allocation "for" loop can be done in $O(n \log m)$. All together, the *Adaptive Fit* can be done in $O(n \log n + n \log m)$, which is equivalent to $O(n \log n)$. □

Reference

1. S. Akoush et al., in *Proc. IEEE MASCOTS*, Predicting the Performance of Virtual Machine Migration. pp. 37–46 (2010)

Chapter 3
Transformation of Data Center Networks

In this chapter, we introduce the *StarCube* framework. Its core concept is the dynamic and cost-effective partitioning of a hierarchical data center network into several star networks and the provisioning of each service with a star network that is consistently independent from other services.

The principal properties guaranteed by our framework include the following:

1. **Non-blocking topology**. Regardless of traffic pattern, the network topology provisioned to each service is non-blocking after and even during reallocation. The data rates of intra-service flows and outbound flows (i.e., those going out of the data centers) are only bounded by the network interface rates.
2. **Multi-tenant isolation**. The topology is isolated for each service, with bandwidth exclusively allocated. The migration process and the workload traffic are also isolated among the services.
3. **Predictable traffic cost**. The per-hop distance of intra-service communications required by each service is satisfied after and even during reallocation.
4. **Efficient resource usage**. The number of links allocated to each service to form a non-blocking topology is the minimum.

3.1 Labeling Network Links

The *StarCube* framework is based on the fat-tree structure [1], which is probably the most discussed data center network structure and supports extremely high network capacity with extensive path diversity between racks. As shown in Fig. 3.1, a k-ary fat-tree network is built from k-port switches and consists of k pods interconnected by $(k/2)^2$ core switches. For each pod, there are two layers of $k/2$ switches, called the edge layer and the aggregation layer, which jointly form a complete bipartite network with $(k/2)^2$ links. Each edge switch is connected to $k/2$ servers through the downlinks, and each aggregation switch is also connected to $k/2$ core switches but

© The Author(s) 2016
L. Tsai and W. Liao, *Virtualized Cloud Data Center Networks:*
Issues in Resource Management, SpringerBriefs in Electrical
and Computer Engineering, DOI 10.1007/978-3-319-32632-0_3

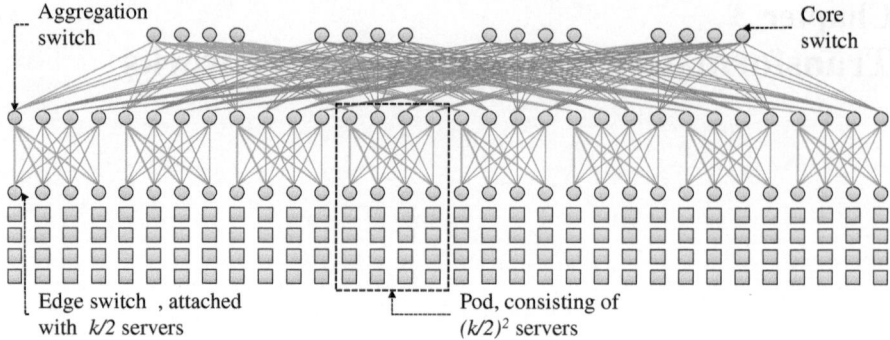

Fig. 3.1 A *k*-ary fat-tree, where *k* = 8

through the uplinks. The core switches are separated into (*k*/2) groups, where the *i*th group is connected to the *i*th aggregation switch in each pod. There are (*k*/2)^2 servers in each pod. All the links and network interfaces on the servers or switches are of the same bandwidth capacity. We assume that every switch supports non-blocking multiplexing, by which the traffic on downlinks and uplinks can be freely multiplexed and the traffic at different ports do not interfere with one another.

For ease of explanation, but without loss of generality, we explicitly label all servers and switches, and then label all network links according to their connections as follows:

1. At the top layer, the link which connects aggregation switch *i* in pod *k* and core switch *j* in group *i* is denoted by $Link_t(i, j, k)$.
2. At the middle layer, the link which connects aggregation switch *i* in pod *k* and edge switch *j* in pod *k* is denoted by $Link_m(i, j, k)$.
3. At the bottom layer, the link which connects server *i* in pod *k* and edge switch *j* in pod *k* is denoted by $Link_b(i, j, k)$.

For example, in Fig. 3.2, the solid lines indicate $Link_t(2, 1, 4)$, $Link_m(2, 1, 4)$ and $Link_b(2, 1, 4)$. This labeling rule also determines the routing paths. Thanks to the

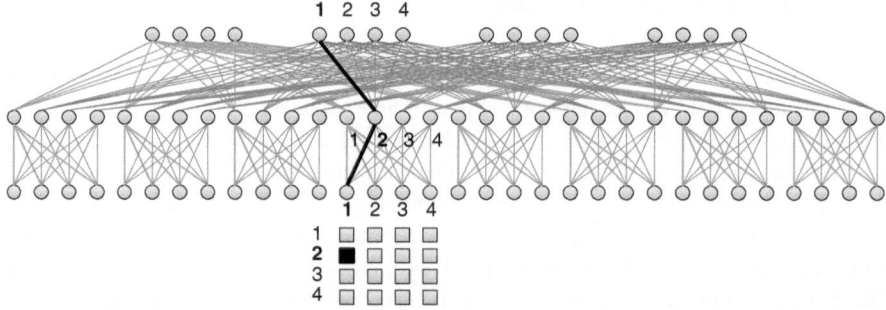

Fig. 3.2 An example of labeling links

symmetry of the fat-tree structure, the same number of servers and aggregation switches are connected to each edge switch and the same number of edge switches and core switches are connected to each aggregation switch. Thus, one can easily verify that each server can be exclusively and exactly paired with one routing path for connecting to the core layer because each downlink can be bijectively paired with one exact uplink.

3.2 Grouping Network Links

Once the allocation of all $Link_m$ has been determined, the allocation of the remaining servers, links and switches can be obtained accordingly. In our framework, each allocated server will be paired with such a routing path for connecting the server to a core switch. Such a server-path pair is called a *resource unit* in this book for ease of explanation, and serves as the basic element of allocations in our framework. Since the resources (e.g. network links and CPU processing power) must be isolated among tenants so as to guarantee their performance, each resource unit will be exclusively allocated to at most one cloud service.

Below, we will describe some fundamental properties of the resource unit. In brief, any two of the resource units are either resource-disjoint or connected with exactly one switch regardless whether they belong to the same pod. The set of resource units in different pods using the same indices i, j is called *MirrorUnits* (i, j) for convenience, which must be connected with exactly one core switch.

Definition 3.1 (*Resource unit*) For a k-ary fat-tree, a set of resources $U = (S, L)$ is called a **resource unit**, where S and L denote the set of servers and links, respectively, if (1) there exist three integers i, j, k such that $L = \{Link_t(i, j, k),$ $Link_m(i, j, k), Link_b(i, j, k)\}$; and (2) for every server s in the fat-tree, $s \in S$ if and only if there exists a link $l \in L$ such that s is connected with l.

Definition 3.2 (*Intersection of resource units*) For any number of resource units $U_1,...,U_n$, where $U_i = (S_i, L_i)$ for all i, the overlapping is defined as $\cap_{i=1...n}U_i = (\cap_{i=1...n}S_i, \cap_{i=1...n}L_i)$.

Lemma 3.1 (Intersection of two resource units) *For any two different resource units $U_1 = (S_1, L_1)$ and $U_2 = (S_2, L_2)$, exact one of the following conditions holds:* (1) $U_1 = U_2$; (2) $L_1 \cap L_2 = S_1 \cap S_2 = \emptyset$.

Proof Let $U_1 = (S_1, L_1)$ and $U_2 = (S_2, L_2)$ be any two different resource units. Suppose $L_1 \cap L_2 \neq \emptyset$ or $S_1 \cap S_2 \neq \emptyset$. By the definitions of the resource unit and the fat-tree, there exists at least one link in $L_1 \cap L_2$, thus implying $L_1 = L_2$ and $S_1 = S_2$. This leads to $U_1 = U_2$, which is contrary to the statement. The proof is done. $\square$

Definition 3.3 (*Single-connected resource units*) Consider any different resource units $U_1,...,U_n$, where $U_i = (S_i, L_i)$ for all i. They are called **single-connected** if there exists exactly one switch x, called the **single point**, that connects every U_i. (i.e., for every L_i, there exists a link $l \in L_i$ such that l is directly connected to x.)

Lemma 3.2 (Single-connected resource units) *For any two different resource units U_1 and U_2, exactly one of the following conditions holds true: (1) U_1 and U_2 are single-connected; (2) U_1 and U_2 do not directly connect to any common switch.*

Proof Consider any two different resource units U_1 and U_2. Suppose U_1 and U_2 directly connect to two or more common switches. By definition, each resource unit has only one edge switch, one aggregation switch and one core switch. Hence all of the switches connecting U_1 and U_2 must be at different layers. By the definition of the fat-tree structure, there exists only one path connecting any two switches at different layers. Thus there exists at least one shared link between U_1 and U_2. It hence implies $U_1 = U_2$ by Lemma 3.1, which is contrary to the statement. The proof is done.   □

Definition 3.4 The set *MirrorUnits(i, j)* is defined as the union of all resource units of which the link set consists of a $Link_m(i, j, k)$, where k is an arbitrary integer.

Lemma 3.3 (Mirror units on the same core) *For any two resource units U_1 and U_2, all of the following are equivalent: (1) $\{U_1, U_2\} \subseteq MirrorUnits(i, j)$ for some i, j; (2) U_1 and U_2 are single-connected and the single point is a core switch.*

Proof We give a bidirectional proof, where for any two resource units U_1 and U_2, the following statements are equivalent. There exist two integers i and j such that $\{U_1, U_2\} \subseteq MirrorUnits(i, j)$. There exists two links $Link_m(i, j, k_a)$ and $Link_m(i, j, k_b)$ in their link sets, respectively. There exists two links $Link_t(i, j, k_a)$ and $Link_t(i, j, k_b)$ in their link sets, respectively. The core switch j in group i connects both U_1 and U_2, and by Lemma 3.2, it is a unique single point of U_1 and U_2.   □

3.3   Formatting Star Networks

To allow intra-service communications for cloud services, we develop some non-blocking allocation structures, based on the combination of resource units, for allocating the services that request n resource units, where n is a non-zero integer less than or equal to the number of downlinks of an edge switch. To provide non-blocking communications and predictable traffic cost (e.g., per-hop distance between servers), each of the non-blocking allocation structures is designed to logically form a star network. Thus, for each service s requesting n resource units, where $n > 1$, the routing path allocated to the service s always passes exactly one switch (i.e., the *single point*), and this switch actually acts as the hub for intra-service communications and also as the central node of the star network. Such a non-blocking star structure is named *n-star* for convenience in this book.

Definition 3.5 A union of n different resource units is called *n-star* if they are single-connected. It is denoted by $A = (S, L)$, where S and L denote the set of servers and links, respectively. The cardinality of A is defined as $|A| = |S|$.

Lemma 3.4 (Non-blocking topology) *For any n-star $A = (S, L)$, A is a non-blocking topology connecting any two servers in S.*

Proof By the definition of *n-star*, any *n-star* must be made of single-connected resource units, and by Definition 3.3, it is a star network topology. Since we assume that all the links and network interfaces on the servers or switches are of the same bandwidth capacity and each switch supports non-blocking multiplexing, it follows that the topology for those servers is non-blocking. □

Lemma 3.5 (Equal hop-distance) *For any n-star $A = (S, L)$, the hop-distance between any two servers in S is equal.*

Proof For any *n-star*, by definition, the servers are single-connected by an edge switch, aggregation switch or core switch, and by the definition of resource unit, the path between each server and the single point must be the shortest path. By the definition of the fat-tree structure, the hop-distance between any two servers in S is equal. □

According to the position of each single point, which may be an edge, aggregation or core switch, *n-star* can further be classified into four types, named type-E, type-A, type-C and type-S for convenience in this book:

Definition 3.6 For any *n-star A*, A is called type-E if $|A| > 1$, and the single point of A is an edge switch.

Definition 3.7 For any *n-star A*, A is called type-A if $|A| > 1$, and the single point of A is an aggregation switch.

Definition 3.8 For any *n-star A*, A is called type-C if $|A| > 1$, and the single point of A is a core switch.

Definition 3.9 For any *n-star A*, A is called type-S if $|A| = 1$.

Figure 3.3 shows some examples of *n-star*, where three independent cloud services (from left to right) are allocated as the type-E, type-A and type-C *n-stars*, respectively. By definitions, the resource is provisioned in different ways:

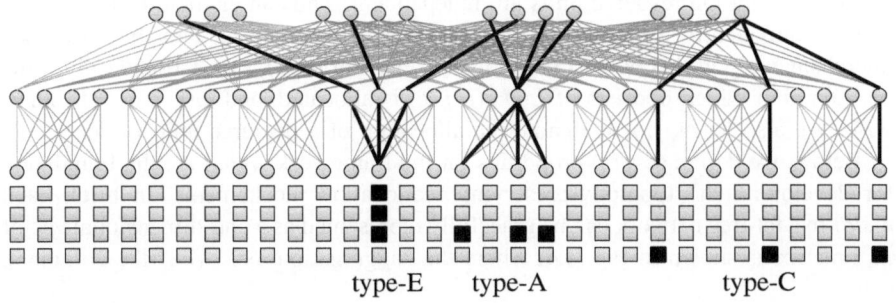

type-E   type-A   type-C

Fig. 3.3 Examples of three *n-stars*

1. Type-E: consists of n servers, one edge switch, n aggregation switches, n core switches and the routing paths for the n servers. Only one rack is occupied.
2. Type-A: consists of n servers, n edge switches, one aggregation switch, n core switches and the routing paths for the n servers. Exactly n racks are occupied.
3. Type-C: consists of n servers, n edge switches, n aggregation switches, one core switch and the routing paths for the n servers. Exactly n racks and n pods are occupied.
4. Type-S: consists of one server, one edge switch, one aggregation switch, one core switch and the routing path for the server. Only one rack is occupied. This type can be dynamically treated as type-E, type-A or type-C, and the single point can be defined accordingly.

These types of *n-star* partition a fat-tree network in different ways. They not only jointly achieve resource efficiency but also provide different quality of service (QoS), such as latency of intra-service communications and fault tolerance for single-rack failure. For example, a cloud service that is extremely sensitive to intra-service communication latency can request a type-E *n-star* so that its servers can be allocated a single rack with the shortest per-hop distance among the servers; an outage-sensitive or low-prioritized service could be allocated a type-A or type-C *n-star* so as to spread the risk among multiple racks or pods. The pricing of resource provisioning may depend not only on the number of requested resource units but also on the type of topology. Depending on different management policies of cloud data centers, the requested type of allocation could also be determined by cloud providers according to the remaining resources.

3.4 Matrix Representation

Using the properties of a resource unit, the fat-tree can be denoted as a matrix. For a pod of the fat-tree, the edge layer, aggregation layer and all the links between them jointly form a bipartite graph, and the allocation of links can hence be equivalently denoted by a two-dimensional matrix. Therefore, for a data center with multiple pods, the entire fat-tree can be denoted by a three-dimensional matrix. By Lemma 3.1, all the resource units are independent. Thus an element of the fat-tree matrix equivalently represents a resource unit in the fat-tree, and they are used interchangeably in this book. Let the matrix element $m(i, j, k) = 1$ if and only if the resource unit which consists of $Link_m(i, j, k)$ is allocated, and $m(i, j, k) = 0$ otherwise. We also let $m_s(i, j, k)$ denote the allocation of a resource unit for service s.

Below, we derive several properties for the framework which are the foundation for developing the topology-preserving reallocation mechanisms. In brief, each *n-star* in a fat-tree network can be gracefully represented as a one-dimensional vector in a matrix as shown in Fig. 3.4, where the "aggregation axis" (i.e., the columns),

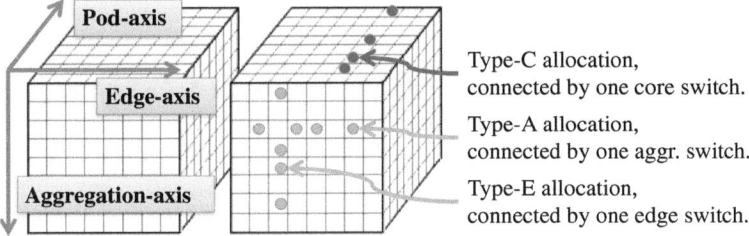

Type-C allocation,
connected by one core switch.

Type-A allocation,
connected by one aggr. switch.

Type-E allocation,
connected by one edge switch.

Fig. 3.4 An example of the matrix representation

the "edge axis" (i.e., the rows) and the "pod axis" are used to indicate the three directions of a vector. The intersection of any two *n-star*s is either an *n-star* or null, and the union of any two *n-stars* remains an *n-star* if they are single-connected. The difference of any two *n-stars* remains an *n-star* if one is included in the other.

Lemma 3.6 (*n-star* as vector) *For any set of resource units A, A is n-star if and only if A forms a one-dimensional vector in a matrix.*

Proof We exhaust all possible *n-star* types of A and give a bidirectional proof for each case. Note that a type-S *n-star* trivially forms a one-dimensional vector, i.e., a single element, in a matrix.

Case 1: For any type-E *n-star* A, by definition, all the resource units of A are connected to exactly one edge switch in a certain pod. By the definition of matrix representation, A forms a one-dimensional vector along the aggregation axis.

Case 2: For any type-A *n-star* A, by definition, all the resource units of A are connected to exactly one aggregation switch in a certain pod. By the definition of matrix representation, A forms a one-dimensional vector along the edge axis.

Case 3: For any type-C *n-star* A, by definition, all the resource units of A are connected to exactly one core switch. By Lemma 3.3 and the definition of matrix representation, A forms a one-dimensional vector along the pod axis.   □

Figure 3.4 shows several examples of resource allocation using the matrix representation. For a type-E service which requests four resource units, $\{m(1, 3, 1),$ $m(4, 3, 1), m(5, 3, 1), m(7, 3, 1)\}$ is one of the feasible allocations, where the service is allocated aggregation switches 1, 4, 5, 7 and edge switch 3 in pod 1. For a type-A service which requests four resource units, $\{m(3, 2, 1), m(3, 4, 1), m(3, 5, 1),$ $m(3, 7, 1)\}$ is one of the feasible allocations, where the service is allocated aggregation switch 3, edge switches 2, 4, 5, 7 in pod 1. For a type-C service which requests four resource units, $\{m(1, 6, 2), m(1, 6, 3), m(1, 6, 5), m(1, 6, 8)\}$ is one of the feasible allocations, where the service is allocated aggregation switch 1, edge switch 6 in pods 2, 3, 5, and 8.

Within a matrix, we further give some essential operations, such as intersection, union and difference, for manipulating *n-star* while ensuring the structure and properties defined above.

Definition 3.10 The **intersection** of two *n-stars* A_1 and A_2, denoted by $A_1 \cap A_2$, is defined as $\{U|U \in A_1$ and $U \in A_2\}$.

Lemma 3.7 (Intersection of n-stars) *For any two n-stars A_1 and A_2, let $A_x = (S_x, L_x)$ be their intersection, exactly one of the following is true: (1) they share at least one common resource unit and A_x is an n-star; (2) $S_x = L_x = \emptyset$. If Case 2 holds, we say A_1 and A_2 are* **independent**.

Proof From Lemma 3.6, every *n-star* forms a one-dimensional vector in the matrix, and only the following cases represent the intersection of any two *n-stars* A_1 and A_2 in a matrix:

Case 1: A_x forms a single element or a one-dimensional vector in the matrix. By Lemma 3.6, both imply that the intersection is an *n-star* and also indicate the resource units shared by A_1 and A_2.

Case 2: A_x is null set. In this case, there is no common resource unit shared by A_1 and A_2. Therefore, for any two resource units $U_1 \in A_1$ and $U_2 \in A_2$, $U_1 \neq U_2$, and by Lemma 3.1, $U_1 \cap U_2$ is a null set. There are no shared links and servers between A_1 and A_2, leading to $S_x = L_x = \emptyset$.   □

Definition 3.11 The **union** of any two *n-stars* A_1 and A_2, denoted by $A_1 \cup A_2$, is defined as $\{U|U \in A_1$ or $U \in A_2\}$.

Lemma 3.8 (Union of n-stars) *For any two n-stars A_1 and A_2, all of the following are equivalent: (1) $A_1 \cup A_2$ is an n-star; (2) $A_1 \cup A_2$ forms a one-dimensional vector in the matrix; and (3) $A_1 \cup A_2$ is single-connected.*

Proof For any two *n-stars* A_1 and A_2, the equivalence between (1) and (2) has been proved by Lemma 3.6, and the equivalence between (1) and (3) has been given by the definition of *n-star*.   □

Definition 3.12 The **difference** of any two *n-stars* A_1 and A_2, denote by $A_1 \backslash A_2$, is defined as the union of $\{U|U \in A_1$ and $U \notin A_2\}$.

Lemma 3.9 (Difference of *n-stars*) *For any two n-stars A_1 and A_2, if $A_2 \subset A_1$, then $A_1 \backslash A_2$ is an n-star.*

Proof By Lemma 3.1, different resource units are resource-independent (i.e., link-disjoint and server-disjoint), and hence removing some resource units from any *n-star* will not influence the remaining resource units.

For any two *n-stars* A_1 and A_2, the definition of $A_1 \backslash A_2$ is equivalent to removing the resource units of A_2 from A_1. It is hence equivalent to a removal of some elements from the one-dimensional vector representing A_1 in the matrix. Since the remaining resource units still form a one-dimensional vector, $A_1 \backslash A_2$ is an *n-star* according to Lemma 3.6.   □

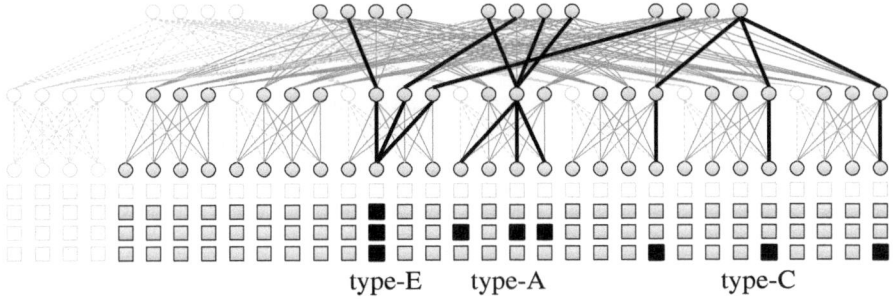

Fig. 3.5 An example of reducing a fat-tree while keeping the symmetry property

3.5 Building Variants of Fat-Tree Networks

The canonical fat-tree structure is considered to have certain limitations in its architecture. These limitations can be mitigated by *StarCube*. As an instance of *StarCube* is equivalent to a fat-tree network, it can be treated as a mechanism to model the trimming and expanding of fat-tree networks. As such, we can easily construct numerous variants of fat-tree networks for scaling purposes while keeping its promising symmetry properties. Therefore, the resource of variants can be allocated and reallocated as of canonical *StarCube*. An example is illustrated in Fig. 3.5, where a reduced fat-tree network is constructed by excluding the green links, the first group of core switches, the first aggregation switch in every pod, the first server in every rack, and the first pod from a canonical 8-ary fat-tree. In this example, a *StarCube* of $4 \times 4 \times 8$ is reduced to $3 \times 4 \times 7$. Following the construction rules of *StarCube*, it is allowed to operate smaller or incomplete fat-tree networks and expand them later, and vice versa. Such flexibility is beneficial to reducing the cost of operating data centers.

3.6 Fault-Tolerant Resource Allocation

This framework supports fault tolerance in an easy, intuitive and resource-efficient way. Operators may reserve extra resources for services, and then quickly recover those services from server failures or link failures while keeping the topologies logically unchanged. Only a small percentage of resources in the data centers are needed to be kept in reserve.

Thanks to the symmetry properties of the topologies allocated by *StarCube*, where for each service the allocated servers are aligned to a certain axis, the complexity of reserving backup resources can be significantly reduced. For any service, all that is needed is to estimate the required number of backup resource units and request a larger star network accordingly, after which any failed resource

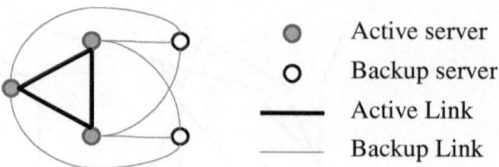

Fig. 3.6 An example of inefficiently reserving backup resources for fault tolerance

units can be completely replaced by any backup resource units. This is because that backup resource units are all leaves (and stems) of a star network and thus inter-changeable in topology. There is absolutely no need to worry about topology-related issues when making services fault-tolerant. This feature is par-ticularly important and resource-efficient when operating services that require fault tolerance and request complex network topologies. Without star network allocation, those services may need a lot of reserved links to connect backup servers and active servers, as shown in Fig. 3.6, an extremely difficult problem in saturated data center networks; otherwise, after failure recovery, the topology will be changed and intra-service communication will be disrupted.

The fault tolerance mechanisms can be much more resource-efficient if only one or few failures may occur at any point in time. Multiple services, even of different types, are allowed to share one or more single resource unit as their backup. An example is shown in Fig. 3.7, where three services of different types share one backup resource unit. Such simple but effective backup sharing mechanisms help raise resource utilization, no matter how complex the topologies requested by services. Even after reallocation (discussed in the next section), it is not required to find new backups for those reallocated services as long as they stay on the same axes. In data centers that are much more prone to failure, services are also allowed to be backed with multiple backup resource units to improve survivability, and those backups can still be shared among services or just dedicated. The ratio of these two types of backups may be determined according to the levels of fault tolerance requested by services or provisioned by data center operators.

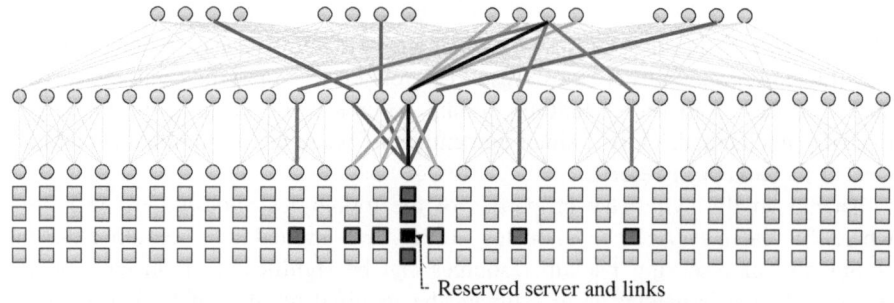

Fig. 3.7 An example of efficiently reserving backup resources for fault tolerance

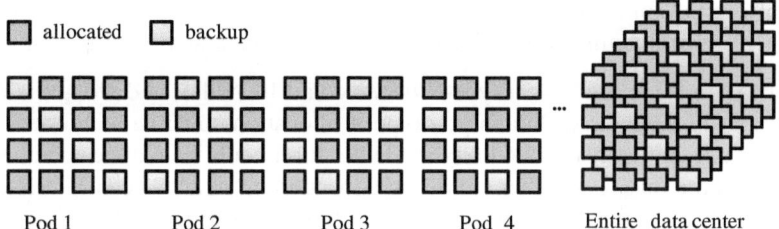

Fig. 3.8 An example of efficiently sharing backup resources for an 8-ary fat-tree

As shown in Fig. 3.8, where there is at least one shared backup resource unit on each axis, fault tolerance can be provided for every resource unit and hence every service, no matter they are type-E, type-A or type-C services. To provide such a property in a data center using a k-ary fat-tree, only $2/k$ resource is required to be reserved. When k equals to 64 (i.e., the fat-tree is constructed with 64-port switches), it takes only about 3 percent of resource to prevent services from being disrupted by any single server failure and link failure.

3.7   Fundamental Properties of Reallocation

Resource efficiency and performance guarantee are two critical issues for a cloud data center. The guarantee of performance is based on the topology consistency of allocations over their entire life cycle. Based on the properties and operations of *n-star*, we design several fundamental reallocation mechanisms which allow an *n-star* to be reallocated while the topology allocated to every service is still guaranteed logically unchanged during and after reallocation. The basic concept of reallocating each resource unit, called an *atomic reallocation*, is reallocating it along the same axes in the matrix, with the reallocated *n-star* remaining a one-dimensional vector in the matrix.

The atomic reallocation promises three desirable properties for manipulating *n-star* allocations. In brief, (1) the topology is guaranteed logically unchanged; (2) the migration path is guaranteed independent (link-disjoint and server-disjoint) to other services or another migration path; and (3) the migration cost is limited and predictable. Using these properties, we can develop some reallocation mechanisms (or programming models) for different objectives of data center resource management.

Definition 3.13 For every *n-star* A or resource unit, the allocation state is either **allocated** or **available** (i.e., not allocated).

Definition 3.14 For any *n-star* A, any pair of *n-stars* (x, z) is called an **atomic reallocation** for A if (1) $A \cup x \cup z$ is single-connected; (2) $x \subset A$; (3) $z \not\subset A$; (4) x is allocated; (5) z is available; and (6) $|x| = |z| = 1$. In addition, **reallocation** is defined as applying the following modifications on resource units in order: $z \leftarrow$ *allocated,* $A \leftarrow (A \cup z)$, migrating service from x to z, $A \leftarrow A\backslash x$, and $x \leftarrow$ *available.*

For example, in Fig. 3.4, a type-E *n-star* could be reallocated along the aggregation axis in the matrix. Similarly, a type-A allocation could be reallocated along the edge axis in the matrix; and a type-C allocation could be reallocated along the pod axis in the matrix. In other words, an allocation on $m(i, j, k)$ can be reallocated to $m(i', j, k)$ if it is type-E, to $m(i, j', k)$ if it is type-A, or to $m(i, j, k')$ if it is type-C.

Lemma 3.10 (Internal topology invariant) *For any n-star A and any atomic reallocation (x, z) for A, A is n-star during and after the reallocation.*

Proof For any *n-star* A and any atomic reallocation (x, z) for A, by Lemma 3.8, $A \cup z$ remains *n-star* during the reallocation. After reallocation, $(A \cup z)\backslash x$ remains *n-star* by Lemma 3.9. Thus the proof is done. $\qquad\square$

Lemma 3.11 (Independent migration path) *For any atomic reallocation (x, z), there exists a migration path P connecting x and z, such that P is in $x \cup z$, link-disjoint and server-disjoint to any n-star A_2 where $x \notin A_2$ and $z \notin A_2$.*

Proof For any atomic reallocation (x, z), let $R = x \cup z$, where R is an *n-star* by the definition and Lemma 3.8 and the definition of atomic reallocation. Further, by the definitions of *n-star* and resource unit, we can find a migration path P by connecting x, the single point of R and z using only the links in R. For any *n-star* A_2 where $x \notin A_2$ and $z \notin A_2$, A_2 is independent to R by Lemma 3.7. Therefore, P is link-disjoint and server-disjoint to A_2. $\qquad\square$

Lemma 3.12 (Independence of atomic reallocations) *For any two atomic reallocations (x_1, z_1) and (x_2, z_2) for some n-stars, where the four n-stars $x_1, z_1, x_2,$ and z_2 are distinct, there exists a path P_1 connecting x_1 and z_1, and a path P_2 connecting x_2 and z_2, such that P_1 and P_2 are link-disjoint and server-disjoint.*

Proof For any two atomic reallocations (x_1, z_1) and (x_2, z_2), where the four *n-stars* $x_1, z_1, x_2,$ and z_2 are distinct, let $R_1 = (x_1, z_1)$ and $R_2 = (x_2, z_2)$, where R_1 and R_2 are both *n-star* by definition and Lemma 3.8. Both R_1 and R_2 are also independent by Lemma 3.7. We can find P_1 by connecting x_1, the single point of R_1 and z_1 using only the links in R_1, and P_2 by connecting x_2, the single point of R_2 and z_2 using the links in R_2. By Lemma 3.11, P_1 and P_2 are link-disjoint and server-disjoint. $\qquad\square$

Lemma 3.13 (Cost of atomic reallocation) *Foy any atomic reallocation (x, z), x is the only migrated resource unit.*

Proof For any atomic reallocation (x, z), by Lemma 3.11, there exists a migration path which is independent of any *n-star* and influences only x and z. Since z is available, x is the only migrated resource unit. $\qquad\square$

3.8   Traffic Redirection and Server Migration

To realize server and path migration in fat-tree networks, some forwarding tables of the switches on the path must be modified accordingly, and some services need to be migrated. The process is different for the various types of *n-star*. Note that type-S may be dynamically treated as any type.

1. Type-E. On each switch, it is assumed that the downlinks and uplinks can be freely multiplexed to route traffic. Therefore, reallocating a type-E *n-star* does not incur real server migration but only a modification in certain forwarding tables. The new routing path will use a different aggregation switch and a different core switch while the allocation of edge switch remains unchanged.
2. Type-A. It invokes an intra-pod server migration and forwarding table modification, by which the path uses a different edge switch and core switch while the aggregation switch remains the same. The entities involved in migration include the current server (as the migration source), the current edge switch, the current aggregation switch, the new edge switch and the new server (as the migration destination). The first two links are currently allocated to the migrating service, and the last two links must not have been allocated to any service. After the migration, the last two links and the link between the current aggregation switch and the new core switch (which is also available) jointly form the new routing path for the migrated flow.
3. Type-C. It invokes a cross-pod sever migration and forwarding path modification. The entities involved in the migration include only the current server (as the source), the current edge switch, the current aggregation switch, the current core switch, the new aggregation switch, the new edge switch and the new server (as the destination). The first three links are being allocated to the service, and the last three links must not have been allocated. After the migration, the last three links of the migration path jointly form the new routing path for the migrated flow.

Reallocating a type-A or type-C allocation requires servers to be migrated among racks or pods. This generally incurs service downtime and may degrade the quality of on-line services. Since such reallocation cost is generally proportional to the number of server migrations, we could simplify it to the number of migrated servers in this book.

Reference

1. M. Al-Fares et al., in *Proc. ACM SIGCOMM*, A Scalable, Commodity Data Center Network Architecture, (2008)

Chapter 4
Allocation of Servers

4.1 Problem Formulation

Based on the transformation of the fat-tree structure, the problem of allocating servers in hierarchical data center networks can be formulated as an integer programming problem. The objective is to maximize resource efficiency and to minimize reallocation cost while satisfying the topology constraints even when some *n-star* allocations are reallocated.

We divide the timeline into epochs. At the end of each epoch, the current allocations and the updated resource requirements of all services are given, and then this integer programming model is used to find the updated allocation for the next epoch. In real-life scenarios, the length of one epoch could depend on the life cycle or dynamic patterns of cloud services. We also assume that any service is allowed to join, scale or leave only at the beginning of an epoch, which are modeled by incrementing or decrementing the number of requested resource units.

Given a matrix M representing a fat-tree and a set of services S, for each service $s \in S$, we denote the requested allocation type by r_s, and the requested number of resource units by d_s. Let r_s be 1, 2 or 3 to represent type-E, type-A and type-C *n-star* allocations, respectively (Table 4.1). A special case is that the allocation could be freely considered as one of type-E, type-A and type-C when $d_s = 1$ (i.e., type-S). We use variable t_s to differentiate the set of topology constraints applied to s when allocating and reallocating s.

$$t_s = \begin{cases} 1 \, or \, 2 \, or \, 3, & \textit{if } d_s = 1; \\ r_s, & \textit{otherwise.} \end{cases}$$

A service is allocated if and only if the exact number of the requested resource units is satisfied and all the allocated resource units jointly form a one-dimensional vector in matrix M. Such topology constraints for each allocated service s are defined as the following two equations, both of which need to be satisfied at the same time.

$$x_s d_s = \begin{cases} \sum_i m_s(i, e_s, p_s), & \text{if } t_s = 1; \\ \sum_j m_s(a_s, j, p_s), & \text{if } t_s = 2; \\ \sum_k m_s(a_s, e_s, k), & \text{if } t_s = 3; \end{cases}$$

$$x_s d_s = \sum_{i,j,k} m_s(i, j, k).$$

If s is allocated in the next epoch, $x_s = 1$; 0 otherwise. For each allocated service s, we let e_s, a_s and p_s denote the indices of the edge axis, the aggregation axis and the pod axis allocated to s, respectively. Let $m_s(i, j, k) = 1$ if the resource unit at (i, j, k) of M is allocated to service s; 0 otherwise. A type-E allocation could be with multiple aggregation switches (indicated by i); a type-A service could be with multiple edge switches (indicated by j); and a type-C allocation could be with multiple pods (indicated by k). For each allocated service s, the index values of these axes must also be constrained by the size of the matrix M. The variables $N^{(e)}$, $N^{(a)}$, $N^{(p)}$ denote the number of edge switches in a pod, the number of aggregation switches in a pod, and the number of pods in the system, respectively.

$$1 \le e_s \le N^{(e)};$$

$$1 \le a_s \le N^{(a)};$$

$$1 \le p_s \le N^{(p)}.$$

The following equation ensures each resource unit in the matrix M could be allocated to at most one service.

$$\sum_s m_s(i, j, k) \le 1.$$

For service s that exists in both the current and the next epochs, by the reallocation mechanism in the framework, the resource units that have been provisioned are only allowed to be reallocated along the original axis in the matrix M. For each service s, we represent its current allocation with variables x_s', p_s', e_s', a_s' and m_s', which have the same meanings as x_s, p_s, e_s, a_s and m_s except in different epochs. For new incoming services, these variables are set to zero. The following equations ensure the topology allocated to each service remains an n-star with exactly the same single point after reallocation.

Table 4.1 Notation summary for the *StarCube* allocation problem

Parameters	
M	The matrix representing the fat-tree
$N^{(e)}$	The number of edge switches in a pod
$N^{(a)}$	The number of aggregation switches in a pod
$N^{(p)}$	The number of pods in the system
S	The set of all services
d_s	The number of resource units requested by service s
r_s	The allocation type requested by service s, where values 1, 2 and 3 represent type-E, type-A and type-C, respectively
x_s'	The flag indicating if service s is allocated in the current epoch
m_s'	$m_s'(i, j, k) = 1$ if the resource unit at (i, j, k) of M is allocated to service s in the current epoch
e_s'	The edge axis where service s is allocated in the current epoch
p_s'	The pod axis where service s is allocated in the current epoch
a_s'	The aggregation axis where service s is allocated in the current epoch
λ_s	The adjustable ratio of profit to cost
Decision variables	
x_s	The flag indicating if service s is allocated in the next epoch
m_s	$m_s(i, j, k) = 1$ if the resource unit at (i, j, k) of M is allocated to service s in the next epoch
e_s	The edge axis where service s is allocated in the next epoch
p_s	The pod axis where service s is allocated in the next epoch
a_s	The aggregation axis where service s is allocated in the next epoch
c_x	The reallocation cost of service s
t_s	The set of topology constraints for service s

$$0 = \begin{cases} x_s x_s' \left(\left| e_s - e_s' \right| + \left| p_s - p_s' \right| \right), & \text{if } t_s = 1; \\ x_s x_s' \left(\left| a_s - a_s' \right| + \left| p_s - p_s' \right| \right), & \text{if } t_s = 2; \\ x_s x_s' \left(\left| a_s - a_s' \right| + \left| e_s - e_s' \right| \right), & \text{if } t_s = 3. \end{cases}$$

The reallocation cost for service s that exists in both current and next epochs is defined as the number of modified resource units.

$$c_x = x_s x_s' \sum_{i,j,k} \frac{\left| m_s(i,j,k) - m_s'(i,j,k) \right|}{2}.$$

The objective function is defined as follows:

$$\max \left\{ \sum_{s \in S} x_s d_s - \sum_{s \in S} \lambda_s c_s \right\}.$$

It is to maximize a utility function so as to aggregate the profit of allocating resource units requested by cloud services and subtract the reallocation cost suffered by the services. The ratio of profit to cost is adjusted by a weighted coefficient λ_s which depends on the profit model of cloud data centers. For cloud services requiring high performance predictability or cloud data centers claiming high availability, λ_s could be relatively large.

Definition 4.1 (*StarCube allocation decision problem, SCADP*) Given a matrix M and a set of services S, not yet allocated resources (i.e., $x_s' = 0$, $\forall s \in S$). The question is: how to allocate all services in S into M?

Theorem 4.1 (NP-completeness) *There is no polynomial time algorithm that solves SCADP, unless P = NP.*

Proof The problem is in NP since we can guess an allocation for all services in S and verify it in polynomial time. The problem can be reduced from the bin packing problem, described as follows: Given a finite set of items U with integer size $t(u)$ for each $u \in U$, partition U into K disjoint sets such that the total size of items in each set is less than or equal to B.

Given any instance of BPP, we can create an instance of SCADP with a polynomial time transformation: (1) for each $u \in U$, create a type-E incoming service with size $t(u)$; and (2) create a matrix M, where $N^{(e)} = K$, $N^{(a)} = B$ and $N^{(p)} = 1$. If SCADP has a solution in which all type-E services are allocated into K edge switches, the items in U can be partitioned into K disjoint sets, and vice versa. Therefore, SCADP is NP-complete and has no polynomial time algorithm unless $P = NP$. □

Solving the problem usually requires intensive computation due to the NP-completeness and the large scale of cloud data centers. For commercial cloud data centers, there are generally several tens of thousands of resource units and numerous dynamic cloud services dynamically joining, leaving or scaling at any time. To efficiently allocate cloud services within an acceptable time and to maximize the profit of cloud service providers, it is needed to develop a polynomial-time algorithm based on the guaranteed properties of the *StarCube* framework to perform online, on-demand and cost-bounded service allocation and reallocation.

4.2 Multi-Step Reallocation

The time complexity of the algorithm and the resource efficiency are the two most critical factors. Instead of exhaustively searching all possible assignments of the matrix, which is impractical for typical data centers, we take advantage of the properties of the framework to reduce the time complexity of searching feasible reallocation schedules. Simply searching all possible atomic reallocations, however, may not be sufficient to make effective reallocation schedules. We use the proved properties of the atomic reallocation, such as the topology invariance and

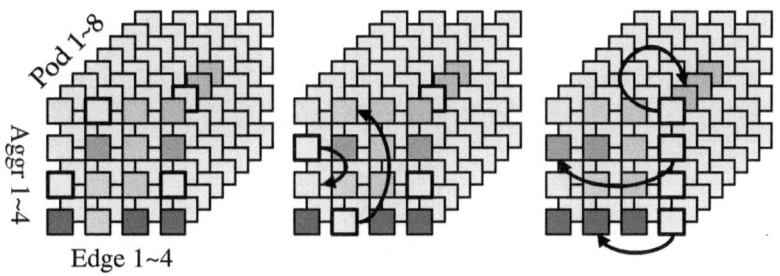

Fig. 4.1 An illustration of the topology-preserving reallocation

independence of atomic reallocations, to combine multiple atomic reallocations into more powerful reallocation schedules while preserving the topology and also ensuring network performance.

We next explain the fundamental concepts of the combined reallocation and demonstrate how to schedule the reallocations. In Fig. 4.1, each matrix represents 8 pods, each with 4 edges and 4 aggregations. The blank squares denote the available resource units while shaded ones denote allocated ones. Distinct services are differentiated by different colors. In this example, there are three independent reallocation schedules; two of them reallocate resource units in a progressive way by combining two atomic reallocations. They first reallocate type-E resource units in the first two columns and then reallocate two type-A resource units in the second and fourth rows. Finally, a service requesting a type-E allocation of four resource units can be allocated in the fourth column.

With the matrix representation, allocating an incoming type-E service is equivalent to searching for sufficient available resource units in any single column, though the available resource units are not required to be next to one another. After performing enough such reallocations, some of the distributed available resource units may be concentrated into one single column so that the aggregated resource units can be allocated to other services. Since every column may have a chance to provide sufficient available resource units after performing some atomic or combined reallocations, the fundamental concept of the reallocation algorithm is to search for as many reallocation opportunities as possible, column by column, and finally select one column to deploy the incoming service.

However, problems still remain in the reallocation process. To reallocate a column, there may be multiple candidate combined reallocations, and they may use the same available resource units, as shown in Fig. 4.2. Such collisions violate the requirements of the *StarCube* reallocation, where any two reallocations are not allowed to share common resource units. To solve this problem, we could exploit the maximum cardinality bipartite matching algorithm [1, 2] to ensure that the selected combined (and atomic) reallocations are resource unit-disjoint. However, this technique works only when the length of the combined reallocations (i.e., the number of involved resource units) is no more than three, because a combined reallocation of longer length may collide not just at the two "ends," for which the

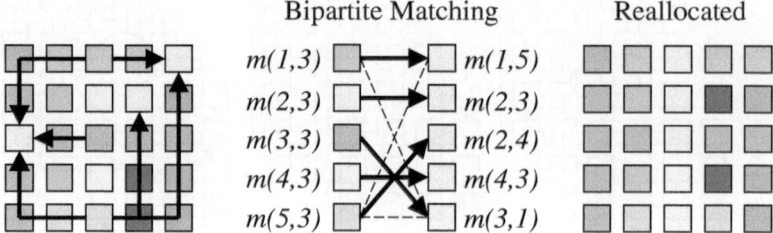

Fig. 4.2 An example of resolving collisions of combined reallocations

matching algorithm is required to apply more than once for a column, leading to higher time complexity. In later sections, we will show that such a limit on the length is sufficient to achieve near-optimal resource efficiency, though longer combined reallocations may render improved efficiency.

In addition to the topology constraints, the length of the combined reallocations may incur a certain cost. To reduce and limit the reallocation cost, we can gradually extend the search space for discovering longer candidate combined reallocations while reserving the possibility of selecting shorter combined reallocations or even atomic reallocations.

4.3   Generality of the Reallocation Mechanisms

We can use the same mechanism to allocate type-E and type-A services by taking advantage of the symmetry of fat-tree networks, and using the transposed matrix (i.e., by inverting the bipartite graph) as the input. With such a transposed matrix, each allocated type-A service is virtually treated as type-E to preserve the topology constraints, and each allocated type-E service is also virtually treated as type-A. The incoming type-A service is treated as type-E and could be allocated to the transposed matrix with the same procedure. Such virtual conversions will not physically affect any service. Because of the longer per-hop distance between servers on type-C allocations, with the algorithm, a service is allocated as type-C only if it cannot be allocated as type-E, type-A or type-S. Such a policy could depend on service requirements, and modifying the procedure for a direct type-C allocation is very straightforward.

4.4   On-Line Algorithm

The proposed algorithm, named *StarCube Allocation Procedure* (SCAP), consists of three sub-procedures. These sub-procedures discover feasible atomic and combined reallocations, and perform a single-pod allocation and a multi-pod allocation.

SCAP tries to allocate an incoming type-E (or type-S) service into a pod. If the service cannot be allocated given that the total available resource units are enough, the network is considered as fragmented and certain sub-procedures will be triggered by SCAP to try to perform reallocations. If, after reallocations, a sufficient number of resource units are available for allocating the incoming service, the reallocations will be physically performed and the service is allocated. Otherwise, SCAP will try to perform multi-pod reallocations if the service prefers to be allocated as type-C rather than be rejected. This is particularly helpful when the available spaces are distributed among pods.

4.5 Listing All Reallocation (LAR)

This procedure is to discover candidate reallocations for each resource unit in a pod. The search scope is limited by a parameter, and we use $L(x)$ to denote the list of discovered candidate reallocations for releasing resource unit x. The detailed steps of the mechanism are shown in Table A.1.

Step 1. If the search scope ≥ 0, discover every dummy reallocation that makes a resource unit available without incurring reallocation and other costs (e.g., $(m(2, 3), m(2, 3))$ in Fig. 4.2). In details, for each available resource unit x, add (x, z) into $L(x)$, where $z = x$.

Step 2. If the search scope ≥ 1, discover every atomic reallocation that makes a resource unit available while incurring exact one reallocation along a row in the matrix (e.g., $(m(1, 3), m(1, 5))$ in Fig. 4.2). In details, for each type-A or type-S resource unit x, add all (x, z) into $L(x)$, where z is an available resource unit in the same row of x.

Step 3. If the search scope ≥ 2, discover every combined reallocation that makes a resource unit available while incurring exact two atomic reallocations (e.g., $(m(1, 3), m(3, 1))$ in Fig. 4.2). In details, for each type-A or type-S resource unit x, add all combined reallocations constructed by (x, y) and (y, z) into $L(x)$, where y satisfies all the following conditions: (1) y is an type-E or type-S resource unit in the same row of x; (2) y is not x; and (3) z is an available resource unit in the same column of y.

Step 4. If the search scope ≥ 3, discover every atomic reallocation that makes a resource unit available while incurring exact one cross-pod reallocation (e.g., $(m(1, 4, 1), m(1, 4, 2))$ in Fig. 4.1). In details, for each type-C resource unit x, add all (x, z) into $L(x)$, where z satisfies all the following conditions: (1) z is an available resource unit; (2) z is in another pod; and (3) z and x belong to the same *MirrorUnits*.

Step 5. Return all $L(x)$ for every x.

4.6 Single-Pod Reallocation (SPR)

This procedure performs reallocations for obtaining n available resource units in a certain column. It first invokes LAR to obtain candidate reallocations, selects some independent reallocations, and then reallocates the target pod such that, in a certain column, there are at least n available resource units.

The procedure repeats multiple iterations with an incremental value, which indicates the search scope of reallocation discovery in LAR and is ranged from 1 to 3 (or up to 2 if type-C is not allowed). Upon finding a successful result, the iteration stops. The detailed steps of the procedure are shown in Table A.2.

Step 1. Use LAR to obtain all cost-bounded candidate reallocations for the matrix.
Step 2. For each column in the matrix, construct a bipartite graph representing all the candidate reallocations (e.g., Fig. 4.2 shows the bipartite graph of the third column), and apply a maximum cardinality bipartite matching algorithm to select resource unit-disjoint reallocations.
Step 3. Select the first column in which there exist at least n unit-disjoint reallocations. Go to the next iteration if such a column cannot be found.
Step 4. For the selected column, select the first n unit-disjoint reallocations derived in Step 2, and reallocate the resource units accordingly.
Step 5. Return the first n available resource units in the selected column.

4.7 Multi-Pod Reallocation (MPR)

This procedure performs reallocations for obtaining n available resource units along the pod axis in the matrix. It first invokes LAR to obtain candidate reallocations for each pod, selects a *MirrorUnits*(i, j), and then reallocates some resource units such that there are at least n available resource units in the selected *MirrorUnits*(i, j).

The procedure repeats multiple iterations with an incremental value, indicating the search scope of the combined reallocations discovery in LAR and is ranged from 0 to 2. Upon finding a successful result, the iteration stops. The detailed steps of the procedure are shown in Table A.3.

Step 1. Use LAR to obtain all cost-bounded candidate reallocations for each matrix.
Step 2. According to the candidate reallocations, for each *MirrorUnits*(i, j), count the number of resource units available (i.e., there exists at least one candidate reallocation.).
Step 3. Select the first *MirrorUnits*(i, j) such that the number derived in the previous step is at least n. Go to the next iteration if such *MirrorUnits*(i, j) cannot be found.

Step 4. In the selected *MirrorUnits*(*i, j*), select and release the first *n* resource units available. For each resource unit, the first candidate reallocation is used in case multiple candidate reallocations exist.

Step 5. Return the first *n* available resource units in the selected *MirrorUnits*(*i, j*).

4.8 StarCube Allocation Procedure (SCAP)

This procedure allocates a type-E service for a request with *n* resource units into the matrix. Note that the same procedure can be used to allocate type-A services, which has been explained before. The detailed steps of the procedure are shown in Table A.4.

Step 1. Select the first pod with the most available resource units.

Step 2. If the number of available resources units is less than *n* in the selected pod, go to Step 5 to perform a cross-pod reallocation (or terminate when type-C allocation is not allowed).

Step 3. In the selected pod, select the first *n* available resource units in the first column with at least *n* available resource units, and, if found, allocate the service to it and the procedure successfully terminates.

Step 4. Invoke SPR in the pod and try to obtain *n* available resource units in the pod, and, if found, allocate the service to it and the procedure successfully terminates.

Step 5. This step is valid only if the service also accepts type-C allocation. Invoke MPR and try to find a *MirrorUnits*(*i, j*) with *n* available resource units, and, if found, allocate the service to it and the procedure successfully terminates.

There are some desirable properties of the proposed mechanism. In summary, (1) every service successfully allocated must be provisioned with an isolated, non-blocking *n-star* topology; (2) the reallocation cost for each service allocation is bounded; (3) the reallocations can be concurrently triggered; and (4) the network performance is consistently guaranteed and isolated among tenants; and the time complexity is polynomial.

4.9 Properties of the Algorithm

Lemma 4.1 (n-star allocation) *In the proposed mechanism, the topology allocated to any service requesting n resource units is an n-star, and also a non-blocking network.*

Proof For any service requesting *n* resource units, a successful allocation is one of Steps 3, 4 and 5 in SCAP. In Steps 3 and 4, the service is allocated into a single column (or a row when the matrix is virtually transposed), and in Step 5, the service is allocated into cross-pod resource units, which must be in the same *MirrorUnits*. The allocation is feasible only if there are n available resource units in such allocation spaces. Therefore, such a feasible allocation must consist of exactly n available resource units connecting to exactly one switch, and by Lemma 3.6, it is an *n-star*. It is also a non-blocking network according to Lemma 3.4.   □

Lemma 4.2 (Number of reallocations) *When allocating any service requesting n resource units by the proposed mechanism, it incurs at most n independent combined reallocations of length two (i.e., each of them consists of at most two atomic reallocations).*

Proof When allocating any service requesting n resource units by the proposed mechanism, the reallocation is equivalent to releasing at most n resource units along a one-dimensional vector in the matrix. With LAR, releasing any resource unit incurs either an atomic reallocation or a combined reallocation which is equivalent to two atomic reallocations (i.e., in a propagation way). SPR and MPR both ensure that any two combined reallocations do not share common resource units. Thus, it incurs at most n combined reallocations, which are independent according to Lemma 3.12, and each of them consists of at most two atomic reallocations.   □

Theorem 4.2 (Concurrency of reallocations) *When allocating any service requesting n resource units by the proposed mechanism, it takes at most two time slots to complete the reallocations, whereas an atomic reallocation takes at most one time slot.*

Proof When allocating any service requesting n resource units by the proposed mechanism, by Lemma 4.2, there are at most n independent combined reallocations of length two. By Lemmas 3.11 and 3.12, the first atomic reallocation of them can be completed in the first time slot, and then the second atomic reallocation of them can be completed in the next time slot. Thus, all reallocations can be completed in at most two time slots.   □

Theorem 4.3 (Bounded reallocation cost) *When allocating any service requesting n resource units by the proposed mechanism, the number of migrated resource units are both bounded by 2n.*

Proof When allocating any service requesting n resource units by the proposed mechanism, by Lemma 4.2, there are at most n independent combined reallocations of length two. By Lemma 3.13, every atomic reallocation migrates one resource unit. Therefore, the number of migrated resource unit is bounded by $2n$.   □

Theorem 4.4 (Multi-tenant isolation) *For any service allocated by the proposed mechanism, the resource units, except the reallocated ones, are consistently and exclusively allocated to the same service for its entire life cycle.*

Proof For any service allocated by the proposed mechanism, by Lemma 4.1, it is allocated with an *n-star*. Since the allocation is formed by available resource units, the *n-stars* allocated to different services are independent according to Lemma 3.7.

By Lemma 3.11, the resource units are also exclusively allocated when other services are reallocated. The proof is done.   □

Theorem 4.5 (Topology consistency) *For any service allocated by the proposed mechanism, the allocation is consistently n-star, and also consistently a non-blocking network.*

Proof For any service allocated by the proposed mechanism, by Lemma 4.1, the allocation is *n-star*, and also a non-blocking network. By Lemmas 3.10 and 3.11, it remains *n-star* during and after reallocation. By Theorem 4.4, it also remains *n-star* when other services are reallocated. Thus it is consistently *n-star*, and also consistently a non-blocking network by Lemma 3.4.   □

Theorem 4.6 (Consistently congestion-free and equal hop-distance) *For any service allocated by the proposed mechanism, any traffic pattern for intra-service communications can be served without network congestion except the servers on reallocation, and the per-hop distance of intra-service communication is consistently equal.*

Proof For any service allocated by the proposed mechanism, by Lemma 3.4, Theorems 4.4 and 4.5, the allocation is consistently an isolated non-blocking network, thus any traffic pattern for intra-service communications can be served without network congestion except for the servers during reallocation, and by Lemma 3.5 the per-hop distance of intra-service communications is consistently equal.   □

Theorem 4.7 (Polynomial-time complexity) *The complexity of allocating any service by the proposed mechanism is $O(N^{3.5})$, where N is the number of servers in a pod.*

Proof The time complexity of the proposed mechanism is dominated by the second step of SPR, which uses a maximum cardinality bipartite matching algorithm to select independent reallocation schedules for each column in the matrix. For each column, we form a bipartite graph for mapping $O(N^{0.5})$ resource units and $O(N)$ reallocation schedules, and hence the bipartite graph has $O(N)$ nodes. With the Hopcroft-Karp algorithm [2], the matching process takes $O(N^{2.5})$ for each bipartite graph with N nodes. There are $O(N^{0.5})$ pods, and $O(N^{0.5})$ columns in each pod. SPR iterates at most three times for extending the search scope in LAR. Thus, the complexity for allocating a service becomes $O(N^{3.5})$.   □

References

1. G.L. Nemhauser, L.A. Wolsey, Integer and combinatorial optimization, John Wiley & Sons, New York, (1988)
2. J.E. Hopcroft, R.M. Karp, An $n^{5/2}$ algorithm for maximum matchings in bipartite graphs, SIAM J. Comp. **2**(4), 225–231, (1973)

Chapter 5
Performance Evaluation

5.1 Settings for Evaluating Server Consolidation

The simulation setup for evaluating *Adaptive Fit* is as follows. The number of VMs in the system varies from 50 to 650. Let the resource requirement for each VM vary in [0, 1] units; the capacity of each server is fixed at one. The requirement of each VM is assigned independently and randomly, and stays fixed in each simulation. The migration cost of each VM varies in [1, 1000] and is independent of the resource requirement. This assignment is reasonable as the migration cost is related to the downtime caused by the corresponding migration, which may vary from a few milliseconds to seconds. The saturation threshold u is assigned with 1, 0.95 and 0.9 so as to demonstrate the ability to balance the tradeoff between migration cost reduction and consolidation effectiveness in different cases.

We use the total migration cost, the average server utilization and the relative total cost (RTC) as the metrics to evaluate the performance of *Adaptive Fit* and compare it with other heuristics. FFD is chosen as the baseline because of its simplicity and good performance in the typical server consolidation problem. Note that since FFD has better performance than FF, we do not show FF in our figures. RTC is defined as the ratio of the total cost incurred in a VM placement sequence F to the maximum possible total cost, namely the maximum migration cost plus the minimum hosting cost. Formally, RTC is defined as follows.

$$RTC = \frac{\alpha \times m + e}{\alpha + 1}$$

$$m = \frac{\sum_{t \in T \setminus \{t_k\}, i \in V, f_j(i) \neq f_{j+1}(i)} c_t(i)}{\sum_{t \in T \setminus \{t_k\}, i \in V} c_t(i)}$$

© The Author(s) 2016
L. Tsai and W. Liao, *Virtualized Cloud Data Center Networks:*
Issues in Resource Management, SpringerBriefs in Electrical
and Computer Engineering, DOI 10.1007/978-3-319-32632-0_5

$$e = \frac{\sum_{t \in T}\left(\left|H'_t\right| \times 1\right)}{\sum_{t \in T, \forall i \in V} r_t(i)}$$

where m is the migration cost of F, which is normalized to maximum migration cost, and e is the hosting cost of F, which is simply defined as the amount of resource allocated normalized to resource requirement. The coefficient α is a normalization ratio of maximum migration cost to minimum hosting cost. It is used to normalize the impact of m and u on the total cost.

For example, consider a system with maximum migration cost of 3 units and minimum hosting cost of 1 unit for the VMs. For a consolidation solution which packs the VMs to servers with total capacity 1.1 times of the total resource requirements (or resource utilization is 90 % on average), and incurs only 0.4 times of the maximum migration cost, its RTC is then given as below, as the migration cost has triple impact on the total cost than the hosting cost.

$$\frac{3 \times 0.4 + 1.1}{3 + 1} = 0.575$$

5.2   Cost of Server Consolidation

The normalized migration cost is shown in Fig. 5.1. It can be seen that *Adaptive Fit* (AF) outperforms FFD in terms of the reduction level of total migration cost, while keeping similar average server utilization levels, as shown in Fig. 5.2. The reduction in total migration cost is stable as the number of VMs increases from 50 to 650. Thus, it demonstrates that our AF solution can work well even for large-scale cloud services and data centers. Besides, by adjusting the saturation threshold u, we see that for AF, the migration cost is decreased as u decreases.

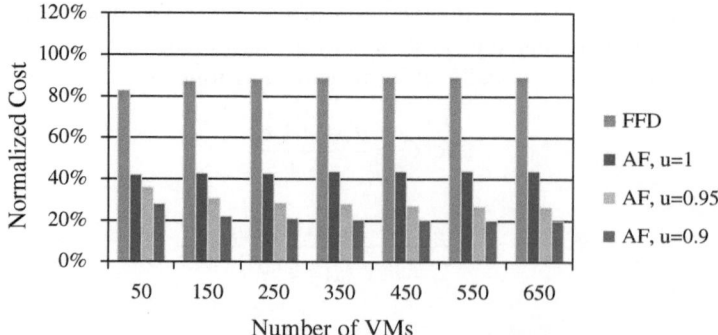

Fig. 5.1 Migration cost of *Adaptive Fit*

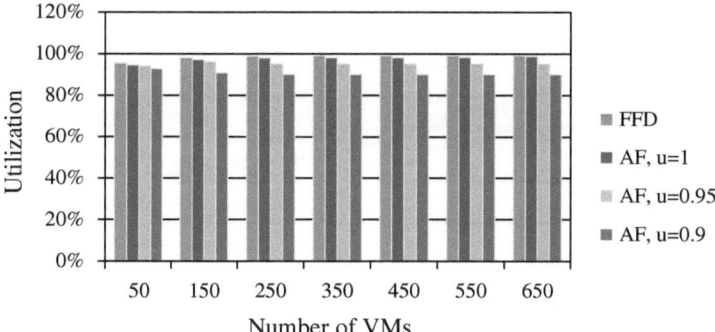

Fig. 5.2 Average utilization of *Adaptive Fit*

When u is decreased from 1 to 0.9, the total migration cost is reduced by up to 50 %. This is because more VMs can be hosted by their last hosting servers, without incurring a migration to disrupt their on-going service.

5.3   Effectiveness of Server Consolidation

Next, we consider the effectiveness of consolidation. Figure 5.2 shows that the average server utilization for AF is very stable and high, with utilization of 97.4 % on average at saturation threshold $u = 1$, which is very close to FFD (98.3 % on average). For a lower u, the need to turn on idle servers for VMs which cannot be allocated to their last hosting servers is more likely, leaving more collective residual capacity in active servers for other VMs to be allocated to their last hosting severs. Therefore, server utilization will be slightly decreased, but migration cost can be significantly reduced. At $u = 0.9$, average utilization is about 90.5 % and average migration cost is further reduced to 21.7 %, as shown in Figs. 5.1 and 5.2.

5.4   Saved Cost of Server Consolidation

To jointly evaluate the benefit and cost overhead of our server consolidation mechanism, we compare the relative total cost RTC caused by AF and FFD. By definition, the total cost depends on different models of revenue and hosting cost; the total cost reduction is shown in Fig. 5.3. We vary the value of α from 0.25 to 32 to capture the behavior of different scenarios. Migration cost dominates the total cost at high values of α, and hosting cost dominates the total cost at low values of α. We fix the number of VMs at 650. As shown in Fig. 5.3, the total cost of AF is much smaller than FFD. FFD incurs very high total cost as it considers only the number of servers in use. The curves of AF match those of FFD very well when α is

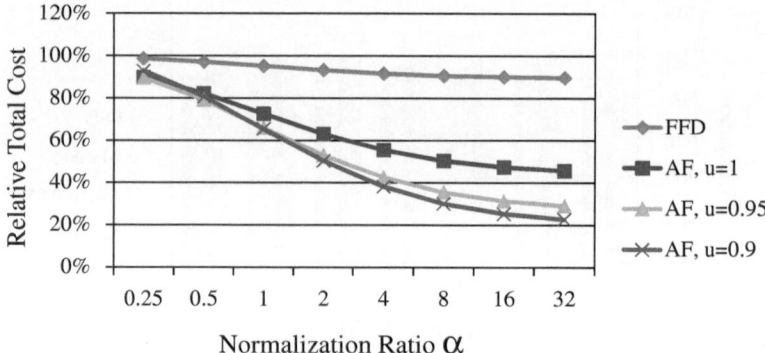

Fig. 5.3 Total cost reduction

very small because then the total cost is dominated by the hosting cost. When α exceeds 0.5, or the maximum migration cost is at least half of the minimum hosting cost, the total cost is much reduced by AF.

In summary, our simulation results show the importance of the adjustable saturation threshold u and the effect of u. (1) For a system with high α, the migration cost dominates the total cost. Therefore, a smaller u will result in enhanced reduction in migration cost and thus lower the total cost. (2) A lower u results in more residual capacity in active servers which can be used to host other VMs without incurring migration. It shows that the solution works well for systems in which low downtime is more critical than high utilization by providing an adjustable saturation threshold to balance the trade-off between downtime and utilization.

5.5   Settings for Evaluating *StarCube*

We also evaluate the performance of mechanisms developed for *StarCube* with extensive experiments. Since this is the first work providing isolated non-blocking topology guarantees for fat-tree networks based cloud data centers, we use a simple allocation mechanism (Method 1) as the baseline, which uses the "first-fit" strategy to perform single-pod allocations, and compare it with the proposed allocation mechanisms (Methods 2 and 3). Method 2 further allows single-pod reallocation. Method 3 further allows type-C allocation and multi-pod reallocation. We implement these methods by modifying SCAP and the sub-procedures. Precisely, Method 1 does not include Steps 4 and 5 of SCAP; Method 2 does not include Step 4 of LAR and Step 5 of SCAP; Method 3 consists of all algorithm steps.

StarCube gives many guaranteed properties, such as consistent isolated non-blocking network allocations. Therefore, it is not needed to evaluate the performance of an individual service, such as task completion time, response time or availability, in the simulation. Rather, we examine the resource efficiency, reallocation cost, scalability, and explore the feasibility for cloud data centers with different dynamic demands. The resource efficiency is defined as the ratio of the total number of allocated resource units to the total number of resource units in the data center. The reallocation cost is normalized as migration ratio (i.e., the ratio of the total number of migrated resource units to the total number of allocated resource units.) For evaluating the scalability, the data center is constructed with a k-ary fat-tree, where k is ranged from 16 to 48 and the number of servers is hence accordingly ranged from 1024 to 27,648 to represent small to large data centers.

In each run of the simulations, a set of independent services is randomly generated. Their requested type of allocation may be type-E or type-A, which is randomly distributed and could be dynamically changed to type-C by Method 3 in some cases mentioned earlier. Each service requests one to N resource units, where N is the capacity (i.e., the maximum number of downlinks) of an aggregation switch or edge switch. The demand generation follows a normal distribution with mean $N/2$ and variance $N/6$ (such that about 99 % of requests belong to $[1, N]$ and any demand larger than N will be dropped). We let the total service demands be exactly equal to the available capacity of the entire data center. In reality, large cloud data centers usually host hundreds and even thousands of independent services. With such a large number, in the simulations we assume the load of services, which is assumed proportional to the number of requested resource units, can be approximated by a normal distribution. We will also show the results based on uniform distribution and discuss the impact of the demand size distribution.

For evaluating the practical capacities for various uses of cloud data centers, we simulate different demand dynamics of a data center. Taking 30 % dynamic as an example, in the first phase, the demands taking 100 % capacity are generated as the input of each allocation mechanism, and then 30 % of the allocated resource units are randomly released. In the second phase, new demands which take the current residual capacity are generated as the input of each allocation mechanism. We collect the data of resource efficiency and reallocation cost after Phase 2. Each data point in every graph is averaged over 50 independent simulation runs.

The simulations for large-scale fully-loaded data centers (i.e., 48-ary and 10 % dynamic) take about 1, 3 and 10 ms in average for Methods 1, 2 and 3, respectively, to allocate an incoming service requesting 10 servers. It shows the run time of the proposed algorithm is of a short delay compared with the typical VM startup time.

5.6   Resource Efficiency of *StarCube*

We evaluate the resource efficiency under different dynamic demands to verify the practicality. As shown in Fig. 5.4, where the data center is constructed with a 48-ary fat-tree (i.e., 27,468 servers), Methods 2 and 3, using the allocation mechanism that cooperates with the proposed reallocation procedures, can achieve almost 100 % resource efficiency regardless how dynamic the demand is. This excellent performance results from rearranging the fragmented resource and hence larger available

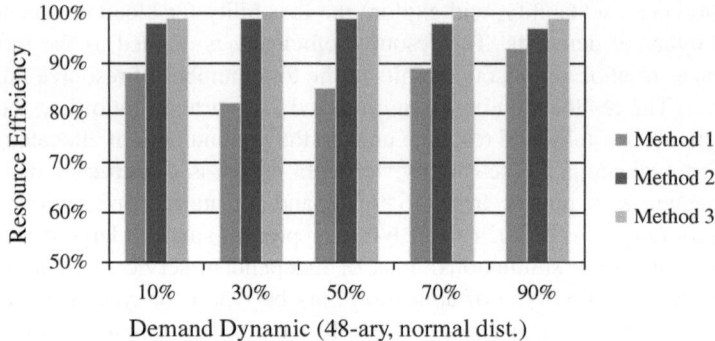

Fig. 5.4 Resource efficiency for various dynamics of demand

non-blocking topologies could be formed to accommodate more incoming service requests which could not be successfully allocated originally. It is shown in the figure that the proposed reallocation mechanisms are feasible to serve diverse dynamics of demands in cloud data centers. The result also shows that even though the proposed framework is based on non-trivial topologies and restricted reallocation mechanisms, near-optimal resource efficiency of data centers is achievable.

Compared with the performance of resource reallocation, the resource efficiency delivered by Method 1 may be degraded to 80 %. The main reason is that the services of dynamic demand may release some resource units at unpredictable positions, fragmenting the resource pool. This makes it harder to find proper available resources for allocating incoming services requesting non-trivial topologies. The problem becomes worse especially when the fragments of residual resource are relatively small at low demand dynamics and the incoming services request more resource units. On the contrary, at higher dynamics, more resources will be released, rendering it more likely to gather larger clusters of available resource units and hence more likely to accommodate the incoming services.

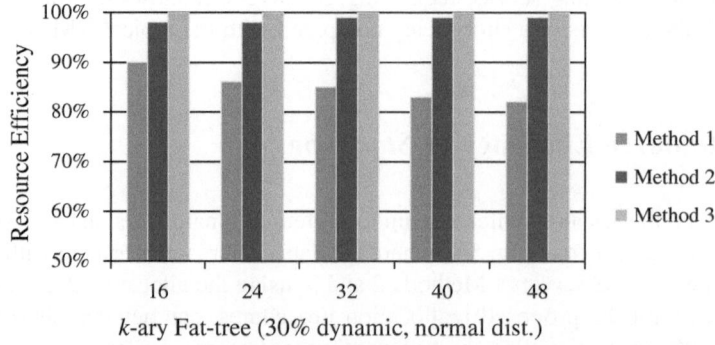

Fig. 5.5 Resource efficiency for various scales

Next, we evaluate the scalability of our mechanism. As shown in Fig. 5.5, where the demand is fixed at 30 %, Methods 2 and 3 can both achieve higher resource efficiency because that the proposed mechanisms effectively reallocate the resources in cloud data centers of any scale. The result shows scalability even in a large commercial cloud data center that hosts more than 20,000 servers. However, since resource fragmentation may occur at any scale and reallocation mechanisms are not supported. Method 1 can only achieve about 80 % resource efficiency.

5.7   Impact of the Size of Partitions

In addition to normal distribution, we also simulate service demands of size uniformly distributed in [1, *N*]. This simulation will support more services than that based on the normal distribution. Because of the nature of capacity limit of a rack or pod in a data center, it is hard to find appropriate spaces to allocate services requesting a large non-blocking topology, particularly in a resource-fragmented fat-tree network. As shown in Figs. 5.6 and 5.7, Method 3 still exhibits consistently better performance in resource efficiency for data centers of various scales or with various dynamic demands.

The better performance derived by Method 3 is thanks to multi-pod reallocation and cross-pod allocation. When a service requesting a large non-blocking network cannot be allocated within a single pod, MPR aggregates the fragmental available resources distributed in multiple pods so as to form an available non-blocking network across these pods. Then, the service is allocated and the resource efficiency is improved. Method 3 suffers from slightly higher computational complexity than Methods 1 and 2 because it considers all pods instead of within one single pod. It also incurs slightly higher communications latency between the servers due to the

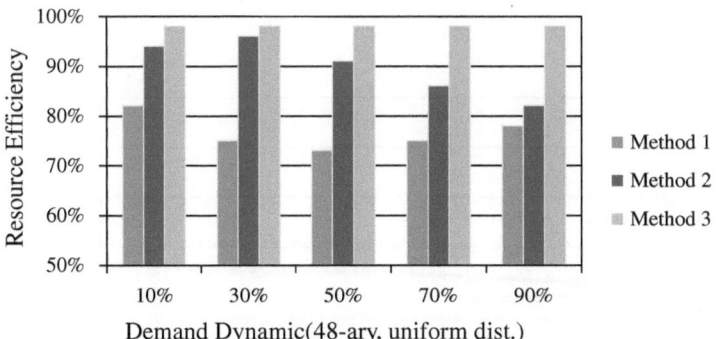

Fig. 5.6 Resource efficiency for various dynamics

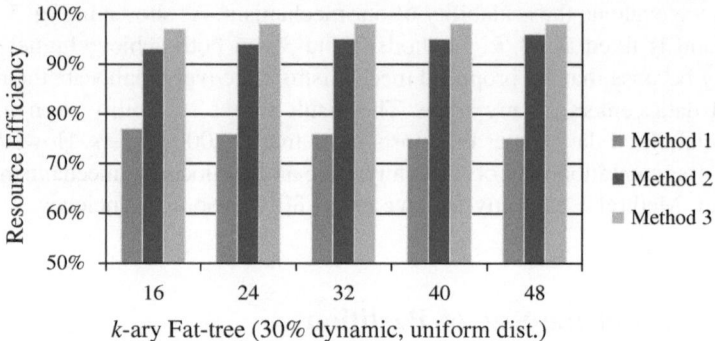

Fig. 5.7 Resource efficiency for various scales

cross-pod allocation. Method 2 leads to good performance when the dynamics of data centers are medium or low. Hence, Method 3 only requires high dynamics (e.g., higher than 70 %).

5.8   Cost of Reallocating Partitions

Inter-rack reallocation and inter-pod reallocation generally incur service downtime and migration time, and their reallocation costs are generally proportional to the number of migrated resource units. We show their results in Figs. 5.8 and 5.9, respectively. Note that, in this framework, every migration is exclusively provisioned with an isolated migration path for migration time minimization, and the number of migrated resource units is also bounded for each service allocation.

As shown in Fig. 5.8, for each resource unit allocation (no matter the type), there are 0.1–0.4 resource units to be reallocated among racks on average. At higher

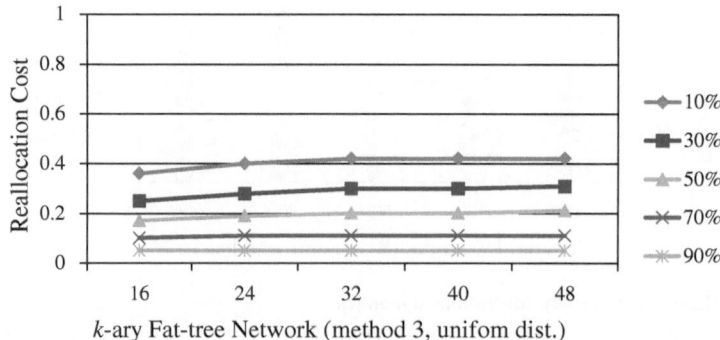

Fig. 5.8 Inter-rack reallocation cost

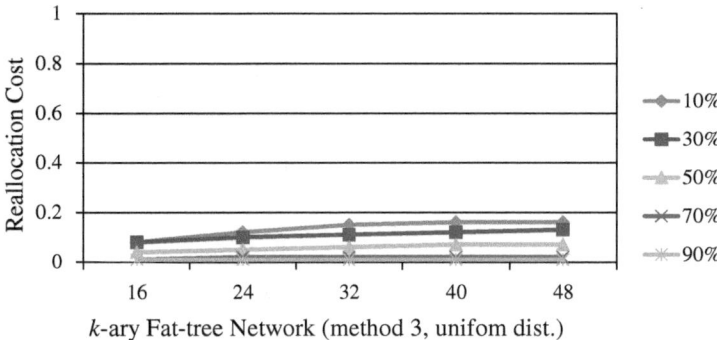

Fig. 5.9 Inter-pod reallocation cost

dynamics, since there are relatively larger clusters of available resource units, more services can be allocated without reallocation and the average cost becomes lower. Even with low dynamics and when the resource pool is fragmented to smaller fragments, the cost is still about 0.4. The inter-pod reallocation cost, as shown in Fig. 5.9, behaves similarly and is smaller than the inter-rack reallocation cost. This is because the proposed mechanism gives higher priority to intra-pod reallocation for reducing cross-pod migration which has longer per-hop distance and may lead to longer migration time. The results show that our method incurs negligible reallocation cost.

Chapter 6
Conclusion

Operating cloud services using the minimum amount of cloud resources is a challenge. The challenge comes from multiple issues, such as the network requirements of intra-service communication, the dynamics of service demand, significant overhead of virtual machine migration, multi-tenant interference, and the complexity of hierarchical cloud data center networks. In this book, we extensively discuss these issues. Furthermore, we introduce a number of resource management mechanisms and optimization models as solutions, which allow the virtual machines for each cloud service to be connected with a consistent non-blocking network while occupying almost the minimum number of servers, even in situations where the demand of services changes over time and the cloud resource is continuously defragmented. Through extensive experiments, we show that these mechanisms make the cloud resource nearly fully utilized while incurring only negligible overhead, and that they are scalable for large systems and suitable for hosting demands of various dynamics. These mechanisms provide lots of promising properties, which jointly form a solid foundation for deploying high-performance computing and large-scale distributed computing applications that require predictable performance in multi-tenant cloud data centers.

© The Author(s) 2016
L. Tsai and W. Liao, *Virtualized Cloud Data Center Networks:*
Issues in Resource Management, SpringerBriefs in Electrical
and Computer Engineering, DOI 10.1007/978-3-319-32632-0_6

Chapter 6
Conclusion

Appendix

See Tables A.1, A.2, A.3, A.4 and A.5.

Table A.1 LAR procedure

```
PROCEDURE: LAR
INPUT: Matrix M (only a pod), search scope s
OUTPUT: Reallocation schedules for every resource unit x in M, denoted
by L(x)


IF s >= 0 THEN
    FOREACH resource unit x IN M
        INSERT  all  schedules  that  release  x  without  actual
        reallocation INTO L(x);
IF s >= 1 THEN
    FOREACH resource unit x IN M
        INSERT  all  schedules  that  release  x  with  a  row-wise
        reallocation INTO L(x);
IF s >= 2 THEN
    FOREACH resource unit x in M
        INSERT  all  schedules  that  release  x  with  a  column-wise
        reallocation followed by a row-wise reallocation INTO L(x);
IF s >= 3 THEN
    FOREACH resource unit x in M
        INSERT  all  schedules  that  release  x  with  a  cross-pod
        reallocation INTO L(x);
RETURN all L(x) in M
```

© The Author(s) 2016
L. Tsai and W. Liao, *Virtualized Cloud Data Center Networks:
Issues in Resource Management*, SpringerBriefs in Electrical
and Computer Engineering, DOI 10.1007/978-3-319-32632-0

Table A.2 SPR procedure

```
PROCEDURE: SPR
INPUT: Matrix M (only a pod), the number of requested space n
OUTPUT: n available resource units in the given pod, null if not found


FOR search scope s <- 1 to 3 // range from 1 to 2 when type-C isn't
allowed
    CALL LAR(M, s) to GET reallocation schedules L(x) for every resource
    unit x in M;
    FOREACH column in M
        SET an empty bipartite graph;
        FOREACH x IN column
            FOREACH schedule IN L(x)
                ADD a link representing the schedule INTO the
                bipartite graph;
        CALL "maximum cardinality bipartite matching" on the bipartite
        graph to GET unit-disjoint reallocations for this column;
    SET selected column AS the first column with at least n unit-disjoint
    reallocations;
    IF selected column == null THEN
        CONTINUE; // expand the search scope
    release the first n releasable resource units of the selected
    column;
    RETURN the first available resource units in the selected column;
RETURN null;
```

Table A.3 MPR procedure

```
PROCEDURE: MPR
INPUT: Matrix M, the number of requested space n
OUTPUT: n available resource units across pods, null if not found

FOR search scope s <- 0 to 2
    CALL LAR(M, s) to GET reallocation schedules L(x) for every resource
    unit x in M;
    FOREACH MirrorUnits(i,j)
        SET Num(i,j) <- the number of resource units available for
        MirrorUnits(i,j);
        // i.e., considering all pods K={k1...kMax}, Num(i,j) =
        |{ m(i,j,k) | k in K, the number of reallocation schedule for
        releasing m(i,j,k) > 0}|
    SET selected MirrorUnits <- the first MirrorUnits(i,j) such that
    Num(i,j) > n;
    IF selected MirrorUnits == null THEN
        CONTINUE; // expand the search scope
    release the first n releasable resource units of the selected
    MirrorUnits;
    RETURN the first available resource units in the selected
    MirrorUnits;
RETURN null;
```

Table A.4 SCAP procedure

```
PROCEDURE: SCAP
INPUT: Matrix M, a service that requests n resource units in a column,
option c, where c == true if cross-pod allocation is allowed, false
otherwise
OUTPUT: TRUE or FALSE, indicating the result of allocation


SET selected pod AS the first pod with the most available resource units;
IF number of available resource units in the selected pod < n THEN
    GOTO TRYCROSS;


SET selected column AS the first column with at least n available
resource units;
IF selected column != null THEN
    allocate the service with the first available resource units in
    the selected column;
    RETURN TRUE;


CALL SPR(matrix of the selected pod, n) to GET available resource units
R;
IF R != null THEN
    allocate service with R;
    RETURN TRUE;
ELSE
    GOTO TRYCROSS;


TRYCROSS:
IF c == false THEN
    RETURN FALSE;
CALL MPR(M, n) to GET available resource units R;
IF R != null THEN
    allocate service with R;
    RETURN TRUE;
RETURN FALSE;
```

Table A.5 *Adaptive fit* procedure

```
PROCEDURE: Adaptive Fit
INPUT: servers H, server capacity y, VMs V, resource requirements of
V in epoch t, placement of V in epoch t-1, saturation threshold u
OUTPUT: placement of V in epoch t


SET A AS ∅; // set of allocated servers
SET A' AS H; // set of unallocated servers
SET V' AS V sorted by resource requirement in decreasing order;


FOR ALL vᵢ ∈ V'
    find the server a ∈ A based on the policy of Worst Fit;
    estimate saturation degree;

    IF the last hosting server l for vᵢ is given THEN
        IF (l is active and sufficient for hosting vᵢ) OR (l is not
        active and saturation degree is higher than threshold u) OR
        (l is not active and there does not exist a to have sufficient
        capacity for vᵢ) THEN
            allocate vᵢ to l;

    IF vᵢ is not allocated to l THEN
        IF (saturation degree is higher than threshold u) OR (there
        does not exist a to have sufficient capacity for vᵢ) THEN
            allocate vᵢ to the first server in A';
        ELSE
            allocate vᵢ to a;

    update fₜ(i), A' and A;
    update the residual capacity of fₜ(i);


RETURN fₜ;
```